**Currents of Change: Impacts of El Niño and La Niña on Climate and Society**

Extreme climatic events such as droughts, hurricanes and floods are spawned worldwide by the periodic warming and cooling of sea surface waters in the central and eastern equatorial Pacific Ocean. The best known of these phenomena is called El Niño, but the equally serious consequences of his lesser known counterpart, La Niña, are now being identified as a result of the 1998–2000 event. These interactions between the ocean and atmosphere are studied by scientists around the world. Although El Niño and La Niña have generally been associated by the media with death and destruction, there are also positive benefits in understanding more about their occurrence, and their global impacts, so that their worst effects can be forecast and mitigated.

This expanded and updated edition of *Currents of Change* explains in simple terms what El Niño and La Niña are, how their effects might be forecast, and what their impacts on humans are throughout the world. Examining for the first time the major El Niño of 1997–98, this book explores what we can learn from past events, what we can do to ameliorate the worst impacts of these extremes, and how climate change might affect El Niño and La Niña events in future decades.

*From the reviews of the first edition:*

"This book is well written, well illustrated and manages to convey a lot of important facts without 'technospeak'. It can be recommended to anyone requiring an introduction to the complexities ... of climate." Elmar R. Reiter, *Meteorology and Atmospheric Physics*.

"Glantz is a passionate advocate of the value of climate research and his book should be compulsory reading for politicians and funding agency heads anywhere. It also provides background material that will be useful to geography and environmental science students and should appeal to anyone interested in the human dimension of climate." J. C. King, *Weather*.

"Michael H. Glantz delivers more than a description of a major climatic phenomenon and its impacts: he presents a case study of science at work, and also shows how to make a book readable." Lothar Lüken, *Earthwatch*.

MICHAEL H. GLANTZ is a Senior Scientist with the US National Center for Atmospheric Research (NCAR) in the Environmental and Societal Impacts Group. He has written and edited several other books including *Drought and Hunger in Africa* (1987), *Drought Follows the Plow* (1994) and *Creeping Environmental Problems and Sustainable Development in the Aral Sea Basin* (1999).

# Currents of Change

## Impacts of El Niño and La Niña on Climate and Society

SECOND EDITION

**Michael H. Glantz**
*Senior Scientist, Environmental and Societal Impacts Group, National Center for Atmospheric Research, Boulder, CO*

PUBLISHED BY THE PRESS SYNDICATE OF THE UNIVERSITY OF CAMBRIDGE
The Pitt Building, Trumpington Street, Cambridge, United Kingdom

CAMBRIDGE UNIVERSITY PRESS
The Edinburgh Building, Cambridge CB2 2RU, UK
40 West 20th Street, New York, NY 10011–444211, USA
10 Stamford Road, Oakleigh, VIC 3166, Australia
Ruiz de Alarcón 13, 28014 Madrid, Spain

http://www.cambridge.org



First published 2001

Printed in the United Kingdom at the University Press, Cambridge

Typeset in Times New Roman 10/12pt [VN]

*A catalogue record for this book is available from the British Library*

*Library of Congress Cataloguing in Publication data*

Glantz, Michael H.
Currents of change: Impacts of El Niño and La Niña on climate and society / Michael H. Glantz – 2nd ed.
p. cm.
Includes bibliographical references and index.
ISBN 0 521 78672 X (pbk.)
1. El Niño Current. 2. Climatic changes. I. Title.
GC296.8.E4 G53 2000
551.47′6 – dc21 00–036291

ISBN 0 521 78672 X paperback

To D. Jan Stewart, El Niño expert, colleague and friend, whose efforts have been tireless to 'get the word out' about El Niño and its influence on societies around the world.

# Contents

*El Niño flip graphics, pp. 123–203.*

*Preface to the second edition* ix
*Acknowledgments to the second edition* xi
*Preface to the first edition* xii
*Acknowledgments to the first edition* xiv

1 Introduction 1

Section I: Emerging interest in El Niño 13

2 El Niño 15
3 A tale of two histories 29

Section II: The life and times of El Niño and La Niña 49

4 The biography of El Niño 51
5 The biography of La Niña 66
6 The 1982–83 El Niño: a case of an anomalous anomaly 84
7 Forecasting El Niño 101
8 Forecasting the 1997–98 El Niño 123
9 Teleconnections 133
10 El Niño's ecological impacts: the Galápagos 146
11 Methods used to identify El Niño 163

Section III: Why care about El Niño and La Niña? 175

12 El Niño and health 177
13 The media, El Niño, and La Niña: a study in media-rology 189

14 Why do ENSO events continue to surprise us? 202
15 What people need to know about El Niño 212
16 Usable science 221

A president's perspective on El Niño, by Alberto Fujimori of the Republic of Peru 225

*Appendix: Chronology of interest in El Niño, by Michael Glantz and Neville Nicholls* 229
*References* 237
*Index* 249

# Preface to the second edition

The 1997–98 El Niño event has challenged the 1982–83 event as the "El Niño of the twentieth century". The most recent El Niño, the last of the second millennium, surprised researchers for at least a couple of reasons. Unexpectedly, it developed very rapidly early in February 1997 and became more intense than anticipated, with sea surface temperatures reaching up to 9 °C in some locations along the Peruvian coast. In May and June of 1998 El Niño decayed at a rate faster than in previous events.

There were other surprises related to the 1997–98 El Niño. A close look at the forecasts produced by various groups suggests that no group forecast in advance the onset of the event or the intensity that it would eventually reach. However, once the observations of changes in sea surface temperatures were recorded, the forecasts were more correct of several of the possible impacts of the event on weather and climate around the globe. Interestingly, this was *the* most watched El Niño ever, with researchers and forecasters using all means available to determine the state of the Pacific Ocean some months in advance: satellites, buoys, ships of opportunity, and computer models. However, most observers held back on their forecasts until the sea surface temperatures began to warm noticeably. This was due in part to the fact that the leading (some say flagship) model for El Niño was projecting that a strong cold event (La Niña) would occur in 1997. It was soon discovered that this model was in error and that the ocean was really in a warming phase, heading quickly toward a major El Niño event.

It seems that El Niño events continue to surprise researchers. Once an event passes, researchers analyze what happened and why, in order to determine why they had missed their forecasts. They then make appropriate adjustments and believe that they have pretty much solved the El Niño puzzle. Some make excuses for their missed forecasts. Still others resort to spin doctoring; that is, presenting their erroneous projections as having been somehow correct. However, the next El Niño often behaves differently from its predecessor; that is, it forms more quickly, it forms at a different time of the year, it has different impacts around the globe, and so researchers and forecasters are surprised once again. And the processes start over, as many researchers set themselves and society to be surprised again.

The graphics in the lower right corner of pp. 123–203 depict sea surface temperatures (in °C). The rectangular box within each graphic is the Niño3.4 region that many researchers use to identify changes in the tropical Pacific Ocean. These pages can be flipped to show the rise and fall of the 1997–98 El Niño and the 1998–2000 La Niña. Flip slowly!

Michael Glantz
*Boulder, Colorado*

The National Center for Atmospheric Research is sponsored by the National Science Foundation.

# Acknowledgments to the second edition

Of course I am still indebted to those who helped me with the first edition of *Currents of Change*. The new addition required help from others. I give special thanks for their help with this edition. Jonathan Patz provided much guidance on climate and health issues. Joey Comeaux did a great job producing the flip charts (pp. 123–203). Mary Zoller learned about ENSO at 90. Tanya Beck and Jan Hopper worked on several drafts. Paul Spyers-Duran had to listen to innumerable details about El Niño. For the rest of my acknowledgments see the following crossword puzzle.

**El niño**

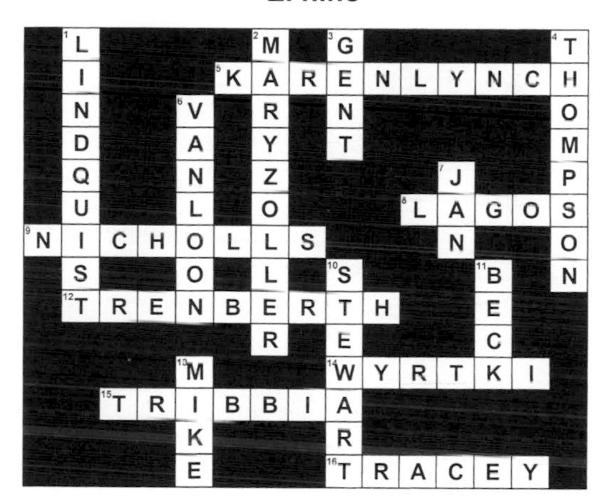


**Across**

5. My wife, Karen, who has been an extremely patient and understanding companion.
8. Pablo Lagos (Geophysics Institute of Peru), who was the first in his country to help me with my first El Niño project in 1976. He is Peru's 'Mr. El Niño.'
9. Neville Nicholls (Australia's Bureau of Meteorology), who has always provided moral as well as scientific support for my El Niño activities. He is *the* El Niño expert.
12. Kevin Trenberth (NCAR), who is always ready to answer my questions.
14. Klaus Wyrtki, whose research inspired many El Niño studies.
15. Joseph Tribbia (NCAR), whose breakfast meetings kept my understanding of science on track.
16. Tracey Sanderson (Cambridge editor), who has had to put up with an author with an attention deficit disorder!

**Down**

1. Jill Lindquist (Philly Junction Diner), who kept my files filled with El Niño cartoons and my cup filled with coffee.
2. Mary Zoller, my mother, who became an El Niño researcher and article clipper at the age of 90.
3. Peter Gent (NCAR), who, like Tribbia, put up with silly questions and gave me the right questions to ask.
4. J. Dana Thompson, a best friend and El Niño mentor.
6. Harry van Loon (NCAR, retired), who is always willing to teach me more about the atmosphere, the Southern Ocean, and ENSO.
7. Jan Hopper (NCAR), a diligent fact-finder who worked without complaint on all drafts of the book.
10. D. Jan Stewart (NCAR), who has carried the brunt of this labor of love. She now knows as much about El Niño as most scholars. Many thanks for 20 years of dedication.
11. Tanya Beck (NCAR), who worked steadily and silently on the various drafts of the manuscript.
13. Mike McPhaden (Pacific Marine Environmental Labs), who is head of the TOGA-TAO Array project.

# Preface to the first edition

More than 20 years ago, when I first visited the National Center for Atmospheric Research (NCAR) in Boulder, Colorado, as a postdoctoral fellow in the Advanced Study Program, I accidentally "stumbled" across El Niño. Here is how it happened. In 1972–73, a major El Niño event occurred off the coast of Peru. That event was linked directly to the collapse of the Peruvian fishing industry, which at that time had been the number 1 fishing industry in the world in terms of the total weight of fish catches. The collapse of the fishery generated considerable scientific interest in the biological impacts of El Niño and, more broadly, in the El Niño phenomenon.

The public attention that was generated by the negative consequences of the El Niño in 1972–73 was short lived, after which societal interest in the phenomenon remained dormant for some years. Only a few governments continued to encourage their scientists to understand the phenomenon better and to determine with greater accuracy how it might directly affect their economies and to forecast when the next such event might occur. The sharpest increase in public awareness about how El Niño events can impact on human activities came in the wake of the 1982–83 event. This has been touted by scientists as the biggest El Niño in a century (biggest can be determined by a variety of factors: the sea surface temperature increase was larger than expected, warmer surface water spanned a larger portion of the Pacific Ocean's surface than in past events, and the impacts on ecosystems and societies were more devastating than during earlier El Niño episodes). The records show that there have been no other events equally as "big" in more than 400 years. This particular event, however, seemed to have captured the attention of the media, especially in North America, as a result of numerous climate-related problems around the world being blamed directly on El Niño influences.

One could convincingly argue that El Niño started to become a household word, more or less, in February 1984, when *National Geographic* magazine chose to present to millions of its readers around the globe a photo essay about the 1982–83 event and its worldwide consequences. El

Niño was even featured on the cover. Later that year, an El Niño article appeared in *The Reader's Digest*, whose subscribers number in the tens of millions and whose editions appear worldwide in more than 18 languages.

Only toward the end of the most recent El Niño event(s) in the 1991–95 period did government agencies in various countries begin to show a much more serious concern about the impacts and applications side of the El Niño phenomenon. Before that time, it was up to the research interests of individual social scientists to "bother" themselves with undertaking research on the societal aspects of El Niño, often unsupported by outside funding. Australia is perhaps *the* leading example of a country that was ready to take El Niño events seriously. Although there had been some research and interest in El Niño events in that country's scientific circles, it was only after the 1982–83 event that Australians sought to look more closely at its impacts on various sectors of their economy, including public safety. In the social sciences, unlike what has happened in the last couple of decades within the physical and biological sciences, no network of researchers with a direct research interest in the social and economic aspects of El Niño events has formed. Yet, El Niño is a natural phenomenon, and improved information about it could yield great benefits to those who choose to use it judiciously. It can lead to more effective decision making.

MICHAEL GLANTZ
*Boulder, Colorado*

# Acknowledgments to the first edition

Many people have helped me with various aspects of the El Niño story. I want to thank them for their assistance and patience in responding to my numerous pleas: Peter Gent, Mark Cane, Joseph Tribbia, Antonio Busalacci, Wang Shao-wu, Tesfaye Haile, Kenneth Mooney, Gary Sharp, Warren Wooster, William Kellogg, George Kiladis, Roger Pielke Jr, Rick Katz, Stephen Zebiak, Claudia Nierenberg, Antonio Magalhães, Dale Jamieson, Elm Sturkol, Carl Hunt, Vicki Holzhauer, Maria Krenz, Brad McLain, Leslie Forehand and Jan Hopper. Their critical reviews were extremely useful. Special thanks go to Michele Betsill for research and administrative support. The lion's share of thanks goes to D. Jan Stewart, who had the onerous task of preparing numerous drafts of the manuscript and editing the text. She exhibited great patience and perseverance while undertaking this task on top of all her other responsibilities, including teaching daring souls how to skydive. I am very much indebted to my Cambridge University Press editor, Tracey Sanderson, who provided me with sorely needed guidance and direction at important junctures in the preparation of this manuscript. My wife, Karen, has been an extremely understanding companion, having given up the summer of 1995 to El Niño! I owe her much more than the summer of 1996.

The National Center for Atmospheric Research is sponsored by the National Science Foundation.

# 1 Introduction

Climate is what you expect.
Weather is what you get.
(Anon)

If you don't like the weather, wait a few days.
If you don't like the climate, move.
(Anon)

## Weather and climate variability

Every year there are extreme climate-related problems around the globe, with droughts occurring in some places and floods, fires or frosts in others. For example, the summer of 1988 was one of a severe drought in the agricultural heartland of North America and extremely low streamflow in the Mississippi River basin. Five years later, in the summer of 1993, a period of very heavy rains led to major flooding along the Upper Mississippi and Lower Missouri rivers and many of their tributaries in the United States Midwest. In the early 1990s, newspaper headlines noted that drought-related food shortages in southern Africa put about 80 million Africans at risk of famine. In August 1992, Hurricane Andrew destroyed southern Florida leaving an estimated US$30 billion in destruction. In early 1995, extreme flooding occurred in western Europe, shaking the confidence of countries such as the Netherlands in their ability to prevent a natural hazard from turning into a national disaster, and challenging their belief that scientific and technological developments had buffered their societies from the consequences of extremely heavy rainfall. This was not unlike the situation in the 1970s and 1980s, when Canadian officials who had sought to "drought-proof" their climate-sensitive agricultural activities in the Prairie Provinces came to realize the impossibility of this daunting task when drought in the region recurred. In 1997–98 the second "El Niño of the century" occurred, only 15 years or so after the first "El Niño of the century".

The point is that record-setting climate events are occurring somewhere in the world each year. In fact, Sir John Houghton, who led a major international program that was designed to assess the level of present understanding of the science of climate change (Houghton *et al.*, 1990, 1996), has suggested that records are being set every year and, if there were a year without such an occurrence, that in itself would be record-setting (J. T. Houghton, cited by Greenpeace International, 1994).

Despite the fact that climate fluctuates on seasonal, annual, decadal, and century and longer time scales, in some years there are many more extreme, though not record-setting, meteorological events and more resulting societal problems than one might expect, such as droughts, floods, frosts, fires, ice storms, or blizzards. One such period was 1972–73, when severe droughts occurred in widely dispersed locations such as Australia and Indonesia, Brazil and Central America, India and in parts of sub-Saharan Africa, and heavy flooding occurred in Kenya, southern Brazil, and parts of Ecuador and Peru. At the time it was suggested that some of these widely dispersed climatic extremes might have had a common geographic origin – changes in sea surface temperatures in the tropical Pacific Ocean (El Niño or EN) and changes in atmospheric pressure at sea level across the Pacific basin (the Southern Oscillation or SO). These combined changes have come to be commonly referred to as El Niño events in the popular media and as ENSO (El Niño–Southern Oscillation) events in much of the scientific literature.

Very briefly, an El Niño event can be described as the appearance from time to time of warm sea surface water in the central and eastern Pacific Ocean near the equator. Folklore suggests that the term "El Niño" (literally, the Spanish phrase for "the Christ Child" or "Baby Jesus") was used by Peruvian sailors and fishermen as a label for the annual appearance of warm water along the western coast of their country by December of each year. In some years, the warming along the coast did not dissipate within the usual few months but lingered for more than a year. This too was called "El Niño". In recent decades, the term "El Niño" has been broadened to include all kinds of anomalous sea surface warming in the equatorial Pacific. Scientists now believe that El Niño events are associated with, if not the cause of, many anomalous devastating weather extremes around the globe.

During the past couple of decades the public has learned of El Niño and its impacts in a sporadic way. It would be mentioned in the popular media only when a big El Niño event was believed to be under way. Many of those articles or news releases were simply reports on the events of the day and were devoid of in-depth discussion of the phenomenon. Once the El Niño event (or threat of it) had passed, the media's interest in it waned rapidly. One of the key reasons for undertaking the preparation of this book was to provide a user-friendly account of what El Niño is, what it does and why we, as members of different societies and professions, need to have more than a passing, intermittent interest in it, an interest that has been limited for the most part to when it occurs every few years or so. All that has changed in the last few years of the twentieth century.

The intense 1997–98 event has served to heighten interest in El Niño to levels never before seen. There has been an explosion of interest in El Niño among the public and, responding to that, the printed and electronic media have produced a few thousand articles on various aspects of the phenomenon and its impacts.

There has also been an explosion in the number of websites on the Internet devoted to El Niño and now its cold counterpart, La Niña. The problem may have shifted from a situation of too little information on El Niño to one of too much. Now we have a situation in which the reader has to be much more discriminating about the El Niño and La Niña information that he or she finds in the media and especially on the Internet.

## El Niño and worldwide climate

The associations or linkages between El Niño events and unusual changes (called anomalies) in normal climate patterns around the globe have been referred to as "teleconnections". These include known, as well as perceived, connections between El Niño events and changes in distant weather or climate-related processes. For example, there appears to be an association between El Niño events and an increased likelihood of droughts in such disparate regions as northeastern Australia, southeastern Africa, Indonesia, the Philippines, northeast Brazil, Central America, and so forth.

One location where ecosystems and human activities are known to bedirectly and, for the most part, adversely affected by El Niño is along the western coast of South America, specifically Ecuador, Peru, and northern Chile. Just about every event, regardless of whether weak or strong, usually has an impact on this region, both on land and in the coastal zone. For example, there is a higher tendency toward torrential rains and flooding during El Niño, in northern Peru and southern Ecuador. Linkages also exist between El Niño events and a below-average number of tropical hurricanes along the east coast of the USA as well as in the location of tropical cyclones off the east coast of Australia, where they tend to shift equatorward by several hundred kilometers. La Niña events, for their part, are associated with an above-average number of hurricanes in the Atlantic and Gulf of Mexico as well as with a westward shift in the tropical cyclone belt in the Pacific.

## El Niño and societal impacts

El Niño is a naturally occurring phenomenon that reappears every few years. To varying degrees, it affects a large portion of the world's population. Potentially useful scientific information about El Niño and its impacts on society will probably go unused, unless there are sustained efforts to educate the public about how to realize the value of seemingly abstract scientific research findings. For this reason alone, it is important for the general public, for managers in various economic sectors, and for policymakers to know more about the El Niño phenomenon, including its teleconnections and their implications for ecosystems and societies around the globe.

The scientific literature and popular media are full of statements about the value to society of being forewarned about the possible onset of an El Niño. On an idealized, abstract level, it is not difficult to find value in forecasts of El Niño events or, for that matter, forecasts of any climate-related environmental change. However, when it comes to a specific El Niño event and its specific impacts in local areas worldwide, it becomes a highly speculative endeavor to place a precise value on such forecasts (Pfaff *et al.*, 1999).

### *El Niño and Peru*

The value to various sectors of Peruvian society of knowing more about El Niño has been mentioned in general statements since at least the end of the 1800s. It is the country that first came up with the name "El Niño" for an oceanic current. Peru, therefore, figures prominently in discussions of El Niño. It was at an International Geographical Congress held in Lima, Peru, in the early 1890s that Peruvian geographer Federico Alfonzo Pezet stated that

> the existence of this counter-current [El Niño] is a known fact, and what is now wanted is that proper and definite studies, surveys, and observations should be undertaken in order to get to the bottom of the question, and find out everything relating to this counter-current, and to the influence which it appears to exercise in the regions where its action is most felt.
> (Pezet, 1895, p. 605)

One of the major regional influences to which Pezet referred was heavy rainfall and flooding in northern Peru that extended well beyond a single season and usually, if not solely, accompanied El Niño events.

At the end of the nineteenth century, one could easily have argued that El Niño was of interest mainly to local populations along the western coast of South America because of the associated disruptions in both normal (i.e., expected) rainfall patterns and agricultural productivity and in the reproductive patterns and availability of fish and bird populations along the coast. Actually, concern was directed not so much toward the apparent adverse impacts on fish as it was toward the bird populations which consumed the small fish that dwelled near the ocean's surface and toward changes in coastal navigation and agriculture.

During major El Niño events, the anchoveta fish population was reduced in number because of increased mortality due to decreased food supply in the warm water. Its number also declined because the fish changed their location, becoming less accessible to fish-eating birds. As a direct result of reduced food supply, millions of adult birds and their chicks perished. The carcasses of thousands of dead sea birds washed up onto Peruvian beaches. In many countries, the occasional high level of mortality among bird populations might receive brief notoriety; not so for the sea birds of Peru (Coker, 1920). Various sea birds earned the name of guano birds because they were highly valued in Peru due to their excrement (called guano). Guano was "discovered" by European chemists in the early 1800s to be rich in nitrogen and phosphorus. Guano was considered to be a valuable export commodity for Peru between 1840 and 1880. It was an excellent fertilizer for agricultural fields in Great Britain, Europe, and the USA. El Niño-related reductions in the guano bird population led to a reduced production of guano in Peru's rookeries on the rocky Chincha Islands and along the rocky coastal areas (Figure 1.1).

### *El Niño and the world*

Because of the advent of manufactured fertilizers, along with other trade-related factors, Peru's ability to export guano waned and, as a result, guano-producing birds and guano production were no longer major generators of Peruvian concern about El Niño or its ecological impacts. Interest in El Niño shifted in the 1950s to the exploitation of the anchoveta population for fishmeal production. With the El Niño-related collapse of that fishery in the early 1970s, the health of fish populations in the eastern equatorial Pacific no longer promoted global interest in El Niño episodes. Today, primary concern centers on the realization that El Niño and La Niña are Pacific basin-wide phenomena that can cause regional perturbations of climate- and weather-related processes throughout most of the world.

Increasingly, the task of understanding El Niño is seen by climatologists and meteorologists as an important key to unlocking mysteries about tropical climate and weather patterns and, to a varying extent, about its influence on regions outside the tropics (called the extra-tropics). In fact, an increasing number of El Niño researchers now claim that they can reliably

*Figure 1.1. The rocky islands along the coast of Peru are the nesting sites for "guano" birds. In the absence of an El Niño event, the birds are highly productive, consuming large quantities of fish, primarily anchoveta, that dwell near the ocean's surface. This is converted into guano, bird droppings that are used as a fertilizer for agriculture. When the warm waters appear in the region, signaling an El Niño (warm) event, the fish become fewer in number and are dispersed, becoming inaccessible to the birds. (Photos courtesy of Jaime Jahncke, Instituto del Mar del Peru.)*

forecast the onset of El Niño. Depending on the particular researcher, claims for lead time (i.e., advanced warning) range from 4 to 12 months. Some of these claims are perceived to be realistic and have captured the attention of policymakers who continue to support physical science research on El Niño and El Niño forecasting efforts in a major way. Disruptions of regional climate patterns and of human activities during El Niño reinforce the need for the development of truly reliable long-range climate forecasts, which can then be used by society to reduce the impacts of climate and weather extremes on itself and on vulnerable ecosystems.

### *El Niño and international science*

By the end of the 1990s, scientific research interest in El Niño had blossomed worldwide. El Niño research is no longer left to Peruvian scientists interested in El Niño's local, adverse ecological consequences for sea birds or fish. Now, they have been joined by the equivalent of a small army of scientific researchers, drawn from all continents and from several academic disciplines, which is actively engaged in individual as well as collaborative research on a wide range of El Niño-related topics. Their shared expectation is to resolve lingering mysteries about the phenomenon and to uncover the underlying mechanisms that perpetuate El Niño and La Niña events and govern the phenomena's life cycles. Such discoveries would probably enable scientists to forecast El Niño and La Niña with a high degree of reliability several months in advance of their onset, peak, and decay phases.

Today, policymakers who have been funding science in general are asking questions about the value to society of their research. With limited national budgets, researchers are having to explain to society in more detail the application and usability of their research findings. Forecasting El Niño and, more broadly, forecasting climate variability from one year to the next, has enormous *potential* benefits to society. Some El Niño research findings can already be used to demonstrate how the scientists can produce "usable science". There has also been a sharp increase in policymaker interest in identifying the environmental and societal aspects of El Niño. The scientific community has only recently come to realize the need for sustained efforts to educate the public, especially policymakers, about the societal importance of this phenomenon.

## El Niño as a "living" thing

Like many other processes in nature, El Niño comes and goes again and again. Like other recurring phenomena such as seasonal vegetation cycles, mountain snowpack, glacial advance and retreat, and sand dune movement, El Niño events wax and wane over time in their responses to global climatic fluctuations.

As with any attempt to discuss a system with many interacting components, identifying the best place to begin is often difficult. For example, how might one best describe the human body? Before discussing the parts of the human body and their functions, one must have an idea of how the body, as a whole, works. By analogy, for El Niño it is necessary to describe the phenomenon (i.e., the body) and then to describe its components and their various interactions. In using such an approach, however, it is difficult to avoid some repetition. This inconvenience aside, the reader will hopefully gain a better picture of El Niño, La Niña, and their impacts on weather and climate anomalies around the globe.

## The word "complex"

The scientific community relies heavily on the use of the term "complex". Atmospheric processes are complex; so too are oceanic processes. El Niño is the result of complex interactions between the atmosphere and the ocean. In reality, though, everything is complex, from the electron that circles the nucleus of an atom to the far reaches of the universe. However, in addition to the acceptance of scientific complexity, the term "complex" has also been used for a variety of other reasons. For example, "complex" has been used as an adjective to suggest that an understanding of the phenomenon under investigation cannot be known in its entirety. It has also been used to suggest that it will take a long time (i.e., a great deal of money) to understand it completely. "Complex" has also been used by scientists as a caveat to note "buyer beware"; that the users of such information should treat it as imperfect information. In some instances, it has been used to suggest that you (the reader) cannot possibly understand all that the scientist could tell you about the phenomenon, so he or she will not bother to try.

Thus, the notion of complexity can be used, on the one hand, to expose the limits to our depth of knowledge or, on the other hand, to hide our ignorance. In this book the term will be used sparingly, assuming that readers are well aware of just how complicated various natural processes and interactions are. How much of the science of El Niño does the non-expert *need* to know? How much detail can be left out or generalized in a description of processes and events, while still conveying an understanding that is correct, even if not complete? That is the challenge of those who seek to write about scientific issues for those of us who are not physical scientists.

## Chapter overview

Chapter 2 presents definitions of El Niño, bringing to light a main source of the confusion that surrounds the phenomenon. It attempts to provide a broad definition for El Niño. Chapter 3, entitled "A tale of two histories," presents a brief history of the emergence and spread of scientific interest in El Niño and in the Southern Oscillation.

The fourth chapter, "Biography of El Niño," describes various characteristics and processes associated with El Niño and the Southern Oscillation, while the fifth chapter presents a biography of La Niña. Chapter 6 discusses the 1982–83 event, which until 1997 had been considered the most intense event in a hundred years. Attention is drawn to the true importance of the all-but-forgotten 1972–73 event.

Chapter 7 considers the value, in theory and in practice, of "Forecasting El Niño," providing a few examples of forecast successes and failures. Also mentioned is the unexpected behavior of air–sea interactions in the equatorial Pacific in the 1991–95 period, as well as the need for an international research institute for climate prediction. Chapter 8 begins with an overview (prepared by Michael McPhaden of NOAA's Pacific Marine Environmental Laboratory) of the evolution of the 1997–98 event, and then summarizes the results of a study that assessed the success or failure of 15 forecasts of that event (Barnston *et al.*, 1999).

Many people are interested in changes in air–sea interactions in the equatorial Pacific, insofar as they believe that their regional climate can be affected by events there. To address this issue, Chapter 9 focuses on the linkages of weather or climate anomalies in distant locations that are believed to be associated with ENSO; i.e., teleconnections. The chapter also raises some questions about how global warming of the atmosphere (climate change) might affect El Niño events. Global warming will probably affect the El Niño process in presently unknown ways. Chapter 10 was written by a researcher who was fortunate enough to have been stationed at the Charles Darwin Research Station in the Galapágos Islands and provides a first-hand, on-site overview of the impacts of the 1997–98 event.

How researchers monitor, investigate and forecast El Niño events is briefly discussed in Chapter 11. In the 1996 edition of *Currents of Change*, some of the major post-war international science programs, starting with the International Geophysical Year (1957–58) were identified and discussed.

Chapter 12 focuses on one of the newest areas of interest in El Niño and La Niña research – their impacts on human health. An increasing number of studies have identified strong connections between the outbreaks of various infectious diseases (e.g., malaria, cholera, dengue fever, Rift Valley fever) and either extreme warm or cold sea surface temperatures in the tropical Pacific. This chapter underscores the urgent need for a rapid improvement in an understanding of climate–health relationships.

Globally, the media took an interest in the 1997–98 El Niño in part because it had been compared by some scientists at its onset to the devastating 1982–83 event. Chapter 13 introduces a brief account of the media's response to this event from the perspectives of a few countries. Also addressed is the issue of "media hype".

Chapter 14 attempts to identify why each El Niño or La Niña seems to surprise scientists, forecasters and the public. With each successive event, researchers learn more about the phenomenon. At the same time, however, researchers realize that the ENSO puzzle may be bigger than they had originally thought. In Chapter 15, I have identified some key aspects of El Niño and La Niña that I think the general public ought to know.

The final chapter addresses the notion of "usable science" as it applies to ENSO research. It encourages governments worldwide to support both ENSO research and the application of research output to societal needs. A chronology of interest in El Niño and La Niña from the mid-1800s to the end of the twentieth century (prepared by Neville Nicholls of the Australia Bureau of Meteorology and myself) is presented in the Appendix.

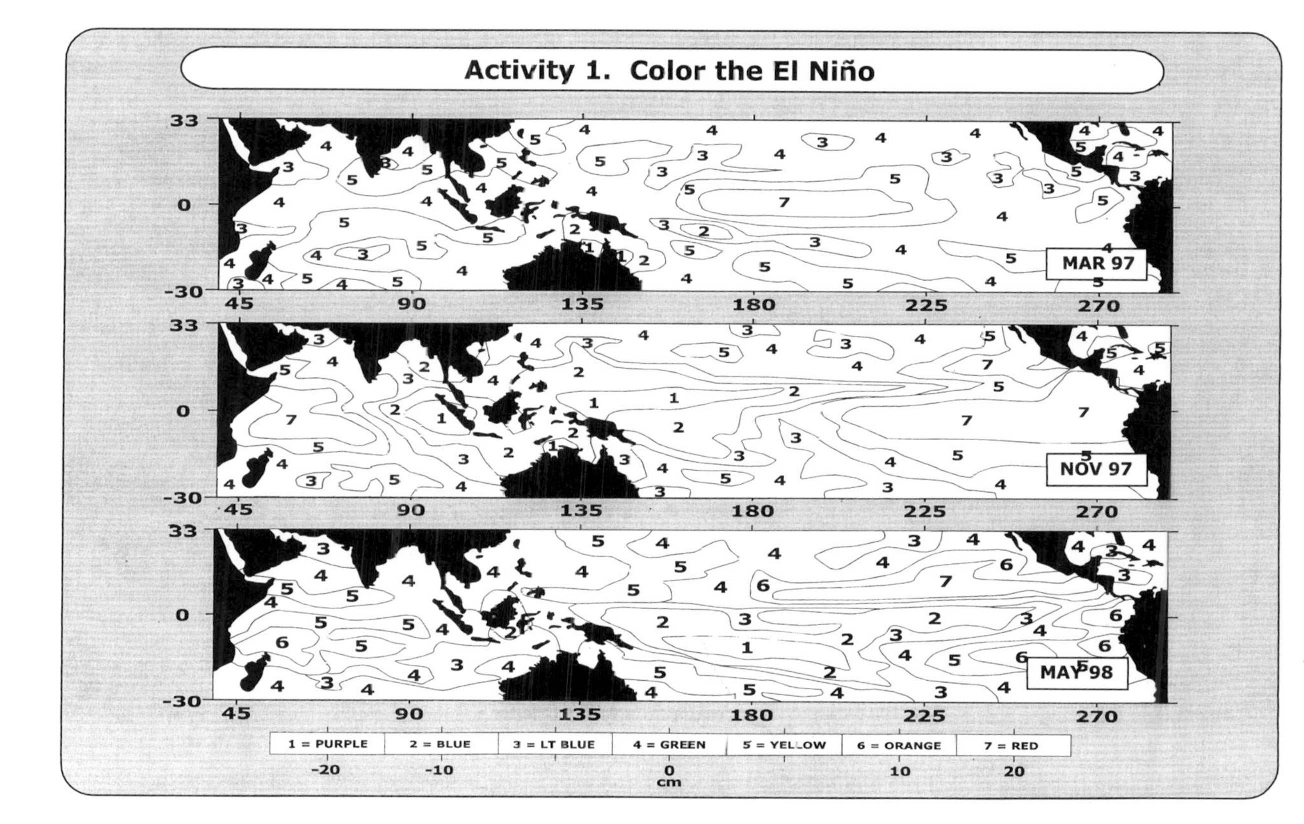


*Color the El Niño. (Material courtesy of NASA/JPL/Caltech and the University of Texas.)*

# Section I
# **Emerging interest in El Niño**

# 2 El Niño

## El Niño definitions

The term "El Niño" means different things to different people. In Spanish, *el niño* means small boy or child. With initial capital letters, El Niño refers to the infant Jesus. To Peruvians, it has an additional meaning: a particular warm ocean current that moves southward along their coast every few years or so. They gave this ocean current the name of El Niño at some time before the beginning of the twentieth century. Although its exact origin and "birthdate" remain unknown, the first time it was noted was in 1892. The popular contemporary version of how it got its name relates to the fact that warm waters appear off the coast of Peru seasonally, beginning around Christmastime (i.e., during the Southern Hemisphere summer, which is the Northern Hemisphere winter). The warm surface water temporarily replaces the usually cold water for a few months along the Peruvian coast. The cold water along the coast is the result of coastal upwelling processes. As a result of such upwelling, deep, cold, nutrient-rich water wells up to the ocean's sunlit surface (called the euphotic zone).

At a Geographical Society Congress in Lima in 1892, Peruvian Navy Captain Camilo Carrillo was apparently the source of information (and rumor) about how the El Niño "current" was named. He made the following statement, which has been repeated many times since then:

> Peruvian sailors from the port of Paita in northern Peru, who frequently navigate along the coast in small crafts, either to the north or to the south of Paita, named this current "El Niño" without doubt because it is most noticeable and felt after Christmas.
>
> (Carrillo, 1892, p. 84)

Occasionally, the warmer water that seasonally appeared off the coast of Peru and Ecuador (a region referred to as the eastern equatorial Pacific) would linger longer than a few months, sometimes lasting well into the following year. These prolonged "invasions" (more correctly, appearances) of warm surface water have led to pronounced disruptions of regional coastal ecosystems involving fish, fish-eating bird populations and economic activities related to fishing and agriculture. El Niño-related torrential rains in northern Peru and southern Ecuador associated with these invasions would occasionally bring devastation to various towns and cities.

In a speech at the International Geographical Society Congress in Lima, Pezet (1895, p. 605) noted "that this hot current has caused the great rainfalls in the rainless regions of Peru appears a fact, as it has been observed that these heavy rains have taken place during the summers of excessive heat". Even though we do not know when El Niño events were recognized as such by various prehistoric cultures along the western coast of South America, we do know that they have occurred over millennia, because the impact of El Niño-related heavy rains and flooding has left its mark (fingerprint) on the natural environment by, for example, leaving flood-related deposits of various soils, rocks and stones in Peru and Ecuador.

As of the beginning of the twentieth century, the connection had not yet been made between El Niño and the various changes in the natural environment around the tropics from the east coast of the African continent to the west coast of South America. Initially, El Niño's impacts, but not necessarily the El Niño phenomenon, were of concern only in Peru and Ecuador, where they were viewed as local manifestations of a local short-term oceanic or atmospheric variation.

By the middle of the 1970s, El Niño had acquired many definitions (Barnett, 1977) and by the late 1990s, several dozen El Niño definitions, ranging from simple to complex, could be found in scientific articles and books (e.g., Glantz and Thompson, 1981; Aceituno, 1992; Trenberth, 1997). The following two definitions serve as examples.

> El Niño: A 12–18 month period during which anomalously warm sea surface temperatures occur in the eastern half of the equatorial Pacific. Moderate or strong El Niño events occur irregularly, about once every 5–6 years or so on average.
>
> (Gray, 1993)

An example of a technical definition that has been used to identify an El Niño event includes the following elements:

> The sea surface temperature (SST) index represents the seasonal SST anomaly (within 4° of the equator from 160° W to the South American coast)... The anomaly [in order to be considered an El Niño] had to be positive for at least three seasons, and be at least 0.5 °C above the average for at least one season, while the SOI [Southern Oscillation Index based on the difference in pressure at sea level between Darwin (Australia) and Tahiti] had to remain negative and below − 1.0 for the same duration.
>
> (Kiladis and van Loon, 1988)

Despite the numerous definitions, some common aspects or characteristics of El Niño do reappear. El Niño

- is an anomalous warming of surface water
- is a warm southward-flowing current off the coast of Peru
- involves sea surface temperature increases in the eastern and central Pacific
- appears off the coasts of Ecuador and northern Peru (sometimes Chile)
- is linked to changes in pressure at sea level (the Southern Oscillation)
- accompanies a slackening of westward-flowing equatorial trade winds
- recurs but not at regular intervals
- returns around Christmastime
- lasts between 12 and 18 months

Some definitions of El Niño also include comments about what an El Niño event does; that is, its impacts, such as "the reduced welling up to the surface of deep cold water", "the appearance of nutrient-poor water", "causes weather changes around most of the globe", and so on.

Periods of cold sea surface temperatures in the central and eastern tropical Pacific are followed by periods of warm sea surface temperatures, which are followed (but not always) by cold or near-average sea surface temperatures, and so on. There are various possible sequences among the three possible states of ocean temperatures: El Niño (warm), La Niña (cold), and average (often referred to as normal). For example, in the early 1990s one El Niño was followed by another, whereas the 1997–98 El Niño was followed by a La Niña event that began in mid-1998. The historical record also shows that an average year can also be followed by another average year and not necessarily by either an extreme warming or cooling of ocean temperatures in the Pacific. Suggestions by various scientists in their articles abound about when an El Niño can be expected to return (called its return period): e.g., 2–10 years, 4–7 years, 3–4 years, 3–7 years, 5–6 years, 5–7 years.

Almost all El Niño definitions today refer to it as an *anomalous* warming of sea surface temperatures off the coast of Peru *and* in the central equatorial Pacific. In a strict statistical sense, perhaps it can be considered as anomalous. It is a change from the seasonal average sea surface temperatures, a departure from an average condition which can be 1, 2, 3, or more degrees Celsius. However, such changes from average conditions are to be expected. From one perspective, an El Niño might be viewed as an anomaly, whereas from a different perspective, it could be considered a normal (i.e., expected) change in sea surface temperatures. This recurring, variable pattern of air–sea interaction that we now call El Niño is part of the normal climate variability and not separate from it.

Clearly, if the scientific community that researches El Niño has shown signs of difficulty in agreeing on a universally accepted definition, should one expect the public and policymakers to be any less confused about the phenomenon? Media coverage of the 1997–98 El Niño, now considered to have been the strongest El Niño of the twentieth century, made matters of definition even more confusing. For example, forecasters claimed success for their forecasts of the onset of an El Niño, when in fact they successfully forecast some of its impacts but not its onset.

After each event, researchers know more than they did before the event occurred. That is a fact. But it seems that each event raises some new unconsidered aspects of El Niño or its impacts. So, although researchers know more in absolute terms, relatively speaking, they know less because the El Niño research problem seems to get bigger.

When the public hears about El Niño from the media, many have a general idea that "something" disruptive is happening in the Pacific Ocean. They may even know that it is happening off the coast of Peru. As scientists continue to discuss El Niño research and theories and the media continue to report them, public knowledge increases by bits and pieces about the interactions between Pacific Ocean sea surface temperatures and the global atmosphere. Today, the public around the globe has been shown that these recurring changes in the equatorial Pacific region have much more far-reaching consequences than merely those that take place along the Peruvian coast. The scientific literature has, since the early 1980s, increasingly referred to the broader Pacific basin-wide changes in sea surface temperature and surface pressure oscillations as ENSO in order to distinguish it from the Peruvian-defined and locally occurring El Niño. ENSO, a term coined in the early 1980s by El Niño researcher Gene Rasmusson, refers to the El Niño–Southern Oscillation phenomenon (Rasmusson 1984a), which is a combination of the interaction which some writers have poetically referred to as "the dance" between ocean temperature changes and atmospheric processes. Interestingly, some researchers have used the terms El Niño and ENSO interchangeably, even within the same scientific article! Other scientists, however, have become dissatisfied with the term ENSO and prefer to use the terms El Niño and La Niña (Philander, 1998).

During the course of the 1990s, but especially as a result of the occurrence of the extraordinary El Niño event in 1997–98, the general public, in many countries, has become familiar with the term El Niño. The use of El Niño is favored by the media. As various media representatives have stated, ENSO is a difficult concept to explain in simple terms to the general public. As a result of the public's initial exposure in the media during the early 1980s to the "newly discovered" mysterious climate-related phenomenon called El Niño, the larger basin-wide changes in the equatorial Pacific are referred to as El Niño.

A dictionary definition of a common word often provides several meanings that have been attributed to that word. Sometimes those meanings are in conflict with each other. Take, for example, the word "leader". One definition notes that a leader has the *ability* to lead; another states that a leader is defined by one who *takes action*. Yet another suggests that a leader is determined by what it says on his/her office door, with no regard for ability. While these competing definitions may overlap, each conveys a different message. Similarly, with respect to El Niño we should accept the fact that the term can have more than a single meaning and that the context in which it is used will help to define its meaning – within that context. Thus, El Niño encompasses both a localized coastal warming of the sea *and* the much broader basin-wide ENSO warm event in the equatorial Pacific. For the reader's sake, "El Niño" is used in this book to describe both the local and basin-wide warming of the sea's surface.

The following box encompasses a wide range of meanings and terms attributed to the El Niño phenomenon by researchers, the media, and the public.

### El Niño

\ 'el nē' nyō *noun* [Spanish] \ 1: The Christ Child 2: the name given by Peruvian sailors to a seasonal, warm southward-moving current along the Peruvian coast < la corriente de El Niño > 3: name given to the occasional return of unusually warm water in the normally cold water [upwelling] region along the Peruvian coast, disrupting local fish and bird populations 4: name given to a Pacific basin-wide increase in both sea surface temperatures in the central and/or eastern equatorial Pacific Ocean and in sea level atmospheric pressure (Southern Oscillation) 5: used interchangeably with ENSO (El Niño Southern Oscillation) which describes the basin-wide changes in air–sea interaction in the equatorial Pacific region 6: ENSO warm event; El Nino [Internet version of spelling]; [Spanish] el fenomeno de El Niño *synonym* warm event *antonym* SEE La Niña \ [Spanish] \ the young girl; cold event; ENSO cold event; non-El Niño year; anti-El Niño or anti-ENSO (pejorative); El Viejo \ 'el vyā hō \ noun [Spanish] \ the old man

Michael H. Glantz

## Measuring El Niño

There are several "types" of El Niño. For example, they can vary in size. The measuring of an event's size can depend on quantitative indicators. One of the most obvious and important indicators is the

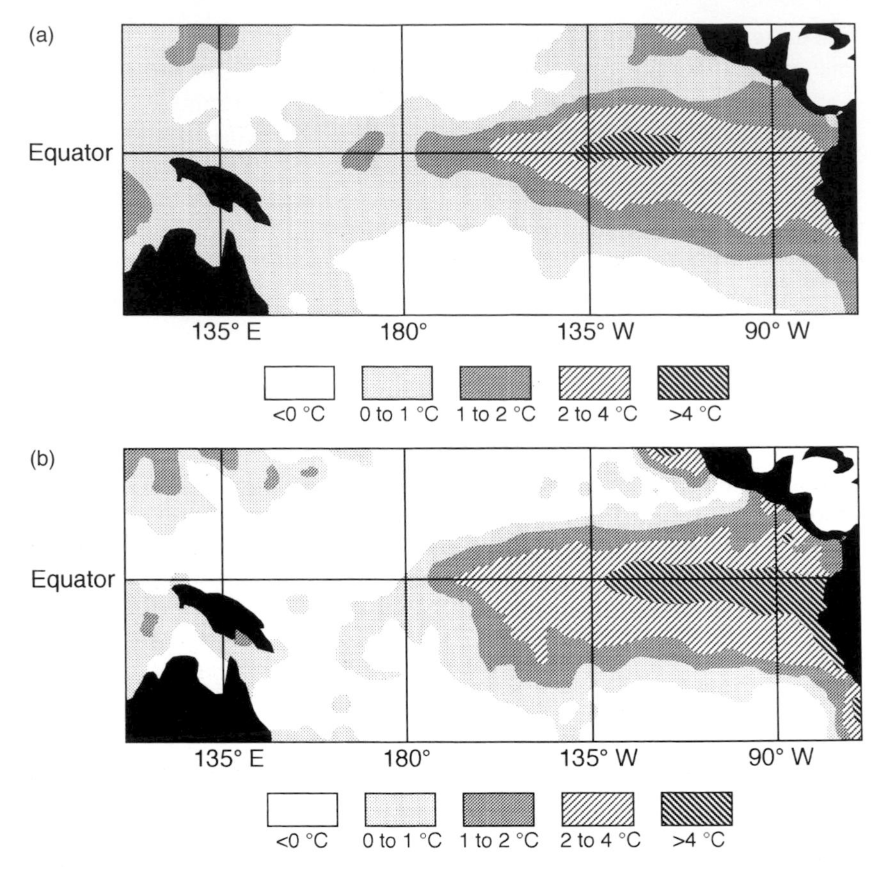


*Figure 2.1. Observed sea surface temperature (SST) anomalies (changes from average conditions) in the Pacific for (a) December 1982–February 1983, and (b) December 1997–February 1998. (Reynolds SST data provided by the NOAA-CIRES Climate Diagnostics Center, Boulder, CO, from their website (www.cdc.noaa.gov).)*

increase in sea surface temperatures in either the central or eastern equatorial Pacific (Figure 2.1a,b).

The larger the above-average increase in temperature, the stronger (more intense) the event. Scientists often refer to the size of an El Niño as being very weak, weak, moderate, strong, or very strong. According to Quinn *et al.* (1987, p. 14 453), "It is assumed that the stronger the event, the greater the amount of damage, destruction, and cost to the nation". They defined these terms by combining physical changes in the ocean with impacts on land, as follows:

> *Very strong events* show extreme amounts of rainfall, flood waters and destruction in Peru, and coastal sea surface temperatures usually reach values of more than 7 °C above normal during some months of the Southern Hemisphere summer and autumn seasons.
>
> *Strong events*, in addition to showing large amounts of rainfall and coastal flooding and significant reports of destruction, exhibit coastal sea surface temperatures in the range of 3–5 °C above-normal range during several months of the Southern Hemisphere summer and autumn seasons.
>
> *Moderate events*, in addition to showing above-normal rainfall and coastal flooding and a lower level of destruction, generally show coastal sea surface temperatures in the range 2–3 °C above normal in the Southern Hemisphere summer and autumn seasons.

Other factors have also been used to determine the "size" of an El Niño. They include the specific location of the warm pool of water in the tropical Pacific, the surface area covered by the anomalously "warm pool" of ocean water, or the depth in the ocean (i.e., volume) of that warm pool. Generally speaking, the larger the surface area and volume of the warm pool, the more intense the event because there is more heat to perturb the atmosphere above it. A weak event is one where the warming is restricted to the western coast of South America, while in a larger event the warm water anomaly would encompass both the central and eastern equatorial Pacific.

Yet another way to measure the size of an El Niño is by how long it lasts. Scientists suggest that El Niño events usually last from 12 to 18 months. Very strong ones rarely last longer than a few years; however, long events have been witnessed a few times in the twentieth century.

In 1991 an El Niño began that some researchers have suggested spanned at least three calendar years, 1991–93. Others have suggested that it did not end in 1993 but continued into early 1995. Was it one long event or a few smaller ones? Scientists still do not agree on how to categorize the changes in sea surface temperatures that occurred in the first half of the 1990s. According to the United States National Weather Service (Halpert *et al.*, 1994), the event that began in 1991 was not the only unusual one in the twentieth century. El Niño events of similar duration occurred in 1911–13 and in 1939–41. While the 1982–83 and the 1997–98 events are considered to have been the most devastating globally in the past century, the 1991 event may prove to have been among the longest. As such, it may have also generated costly impacts, but no research has attempted to cost its damages or benefits. Perhaps another category should be added to describe the size of El Niño – "*extraordinary*". Such a new category seems to be more relevant now than it did in 1990. Scientists labeled the 1982–83 event the "El Niño of the century". They then labeled the 1997–98 event the "El Niño of the century". Now there are two "El Niño events of the twentieth century" in the history books. As we start a new century, it is better to move away from the "...of the century" notion toward a generic category of extraordinary, however researchers choose to measure it.

## Determining an El Niño's size through its impacts

Determining El Niño's size often involves subjective decisionmaking which is often presented as scientific fact. In the late 1970s, William Quinn and his colleagues (Quinn *et al.*, 1978) identified and categorized El Niño events going as far back in history as the early 1500s. To compile their El Niño events, they relied at first on the observations made by historical records of El Niño events and their direct impacts in the twentieth century. To identify El Niño events in earlier times, where past episodes had not been directly observed, they relied on so-called proxy (i.e., indirect) information in various Pacific Rim countries or information that could serve as indirect indicators of El Niño. Researchers have gathered proxy information about rainfall and ocean temperatures from a variety of sources, including the personal diaries of travelers in the region, records of the large number of dead guano birds along the Peruvian coastline, guano-mining records, fish catches of warm water fish species that appear during El Niño along Peru's coast, plantation harvest records in Indonesia, ships' logs, and physical and cultural evidence of floods and mudslides that took place centuries ago (Quinn *et al.*, 1987). These researchers also ascribed intensity to the El Niño events that they uncovered.

In addition to examining environmental changes linked to an El Niño in order to determine its intensity, one can consider El Niño's impacts on human societies as a measure. Some El Niño events have had only local or regional tropical Pacific impacts on society through, for example, heavy rains, flooding, and mudslides that have destroyed villages as well as the transportation infrastructure between villages. Tim Flannery (1995) in his book *The Future Eaters* suggested how El Niño events over millennia have affected climatic, ecological and social development on the Australian continent. Stronger events have major negative effects on crop yields and food production at great distances from the equatorial Pacific region such as in Zimbabwe in southern Africa or northeastern Brazil in South America.

El Niño's impacts on certain ecosystems can provide information on its intensity. Various ecological changes worldwide coincided with the 1982–83 El Niño (Hansen, 1990). For example, Coffroth and colleagues (1990, p. 141) observed that "1982–83 witnessed the most widespread coral bleaching and mortality in recorded history". Destruction of coral reefs around the globe has also raised the specter of possible impacts of global warming on natural marine ecosystems. Very strong El Niño events can provide a glimpse of potentially devastating higher sea surface temperatures impacts on coral. Thus the larger the El Niño, the more likely it is that there will be both an increase in the number of remote locations that are affected by it and an increase in the total cost of all of its adverse impacts. Figure 2.2 provides the reader with a glimpse of the impacts of the extraordinary (very intense) 1997–98 El Niño event.

## A note of caution

Societies change over time in many ways. For example, population numbers increase and there are shifts or expansions in the location of where greater numbers of people live. The societal use (or lack of use) of new technologies and techniques (i.e., ways of doing things) can alter the risks associated with the impacts of natural hazards. All other things being equal, changes that occur in society, many of which are the result of human decisions, can alter (for better or for worse) the degree of vulnerability of societies in the face of climate-related natural hazards such as El Niño. Not all natural hazards become natural disasters.

An example of a shift in a climate-related risk would be the result of the population movement over the past several decades in the USA toward the Atlantic and Gulf coasts. A hurricane making landfall along the US coastline today – having the same intensity and magnitude of one a few decades ago – would probably cause much greater damage to property and life because of the large increase in the population concentrations in coastal settlements. So, when seeking to measure the intensity of an El Niño event in this situation, it would be very misleading to blame El Niño for the "damage" that it might leave in its wake. Part of that damage (or, stated in another way, the increase in the risk of damage) can be directly attributable both to government and private decisionmakers who permitted, if not encouraged, an increasing number of people to move into harm's way. Thus the severity as well as the cause(s) of impacts blamed on El Niño events must be assessed with great care, before one attributes those impacts with confidence to an El Niño that happened to take place at the same time as unrelated demographic changes.

## El Niño as a natural hazard

A community of researchers is focused on natural hazards. For the most part, such hazards include quick-onset events: river flooding, blizzards, avalanches, tsunamis, earthquakes, and hurricanes.

Ian Burton and his colleagues (1993, pp. 35–6) have listed general characteristics that they believe can best be used to define a hazardous event: magnitude, frequency, duration, areal extent, speed of onset, spatial

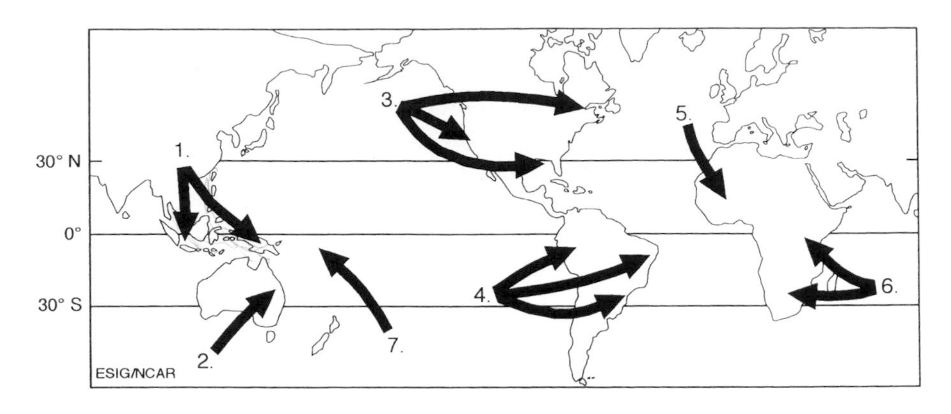


*Figure 2.2. Examples of the impacts of the 1997–98 El Niño on regional climate, ecosystems, and on human activities.*

**Arrow 1.** *Indonesia was plagued with drought and widespread forest fires, many of which had been purposely set by companies as a way to clear land illegally in order to plant plantation crops such as for palm oil production. The fires, however, were blamed on El Niño. El Niño's adverse impacts on agricultural production and the dwindling availability of water resources occurred at a difficult political and economic period in Indonesia's history. The Suharto government had been deposed and new elections called for. Indonesia was also confronted by an independence movement in East Timor. Severe El Niño-related frost and drought in Papua New Guinea led to food shortages that contributed to the only famine to have occurred during the 1997–98 El Niño.*

**Arrow 2**. *Severe impacts of an intense El Niño were forecast for various parts of eastern and northern Australia. As a result of the forecast and of worsening drought conditions, the Australian government reduced its projections of agricultural output and its national economic projections. Farmers and herders used the forecast of El Niño-related drought (and bush fires) to alter their production activities accordingly. However, in the midst of drought, 3 weeks or so of rains favorable to agriculture and rangelands occurred. The government once again revised its economic projections upwards. One Australian newspaper headline referred to this timely wet spell in the midst of severe meteorological drought as "the billion-dollar rains".*

**Arrow 3**. *California was deluged with extremely heavy rains in early 1998. Such a 300–400% increase in rainfall had been projected for late 1997 and, when it did not occur then, the media and others criticized the El Niño forecasters. They referred to the forecast of El Niño as "El No-Show". The rains were late but they did come in the amounts forecast. The US Gulf states were also affected by an anomalous cool* *and wet winter, as expected. There were fewer Atlantic hurricanes than normal, also as expected. The northern tier of the USA witnessed a relatively mild warm winter while the southern tier had a cooler and wetter winter. The drought forecast for the Pacific Northwest failed to occur, as it was a near-normal winter. Killer tornadoes in Alabama and in central Florida were blamed on El Niño, rightly or wrongly, as were the devastating ice storms in the northeastern part of North America (upper New England and Quebec).*

**Arrow 4**. *For the first time in history the governments of Peru and Ecuador took a pro-active response to the forecast in mid-1997 of the impacts of an impending intense El Niño about 6 months into the future. World Bank loans were provided to these countries enabling them to more effectively mitigate the expected impacts of torrential rains and flooding. Nevertheless, despite such pro-active efforts, damage to life, infrastructure and property was still of major proportions. Northeast Brazil was adversely affected by drought, while southern Brazil, Uruguay and northern Argentina witnessed favorable to heavy rains and flooding. Flooding followed a few years of drought in central Chile with both positive and negative effects; refilling empty reservoirs and destruction of property, respectively.*

**Arrow 5**. *No direct linkages were suggested between the climate conditions in the West African Sahel and the occurrence of the intense El Niño in 1997–98.*

**Arrow 6**. *The forecast of the impacts of the 1997–98 El Niño called for drought throughout southern Africa. Governments in the region were told to take appropriate actions in the light of likely El Niño-related food production problems, especially food shortages. The drought in southern Africa, however, did not occur. Heavy rains appeared in East Africa with Kenya receiving the brunt of the damage: destruction of its infrastructure as a result of severe flooding and death as a result of an outbreak of Rift Valley Fever and a sharp increase in the number of malaria cases. This unexpected El Niño-related situation in eastern and southern Africa was probably the result of the unusually warm sea surface in the western Indian Ocean along the east coast of the African continent.*

**Arrow 7.** *Most of the islands in the South Pacific suffered from drought and severe water shortages as a result of the 1997–98 El Niño event. More specifically, water shortages occurred in Fiji, the largest of the islands in this region, Tuvalu, Tonga, the Marshall Islands, the Solomons, among others including one of the smallest of these island nations, Nauru. French Polynesia was hit by a record-setting number of cyclones. Kiribati (pronounced kiri-bahs) was the only Pacific island nation that received good rains during El Niño.*

dispersion, and temporal spacing. They defined each of these characteristics as follows:

- *magnitude*: only those occurrences that exceed some common level of magnitude are extreme
- *frequency*: how often an event of a given magnitude may be expected to occur in the long-run average
- *duration*: the length of time over which a hazardous event persists, the onset to peak period
- *areal extent*: the space covered by the hazardous event
- *speed of onset*: the length of time between the first appearance of an event and its peak
- *spatial dispersion*: the pattern of distribution over the space in which its [impacts] can occur
- *temporal spacing*: the sequencing of events, ranging along a continuum from random to periodic

El Niño is not yet considered such a hazard, although it meets most, if not all, of the criteria used to describe natural hazards. The *magnitude* of an El Niño is defined by the size of the departure from the long-term average of anomalously warm sea surface temperatures in the central and eastern Pacific. *Frequency* relates to its return period, which scientists have suggested is of the order of 2 to 10 years (researchers once suggested that a strong El Niño occurs every 8 to 11 years on average, and a weak to moderate one every 2 or 3 years). The *duration* of El Niño events is generally of the order of 12 to 18 months, although there have been some notable exceptions. The *areal extent* could be interpreted to mean the worldwide spatial extent of the impacts of El Niño and its teleconnections. This would vary with the severity of the event, with major El Niño events being linked to major worldwide impacts and minor ones linked to localized or regional impacts across the tropical Pacific. The *speed of onset* of El Niño is of the order of one or more months. The onset of the 1997–98 events developed at a surprisingly rapid rate in February 1997. *Spatial dispersion* refers to the area in the central and eastern Pacific that is encompassed by the anomalously warm sea surface temperatures. *Temporal spacing*, with respect to El Niño, refers to the return period, which, on average, is 4.5 years with a range of 2 to 10 years.

As one can see, these characteristics apply well to El Niño. Because the general characteristics of El Niño can be encompassed by the general criteria used to define a natural hazard, I believe that the El Niño phenomenon merits inclusion in the official list of natural hazards. Such an explicit designation could help to increase the level of research on El Niño's societal aspects.

## El Niño and La Niña as hazard spawners

Clearly, El Niño and La Niña are parts of a natural process that have been associated with various kinds of hazard. It is only within the last three decades that El Niño was discovered to have such adverse impacts on climate-related processes. The worldwide hazards spawned by these phenomena include droughts, floods, frosts, fires, landslides, and infectious disease outbreaks. Perhaps if we link the originator of the natural hazards (e.g., El Niño) more closely to its resulting hazards, societies can shift their short-term (tactical) responses toward prevention and mitigation (pro-action) and away from a reliance on reaction (i.e., adaptation and clean-up). This would "push" the early warning of specific El Niño-related hazards further "upstream," thereby providing more lead time for societal coping mechanisms to come into play. This would also make El Niño forecasts *the earliest warning* of potential climate-related problems (e.g., food, energy, water, and public safety).

The hazards research community has opposed considering El Niño as a hazard, even though El Niño's characteristics meet the definition of a natural hazard. Some argue that, like winter, it just is a condition of the natural environment. Blizzards or ice storms, for example, happen in wintertime, and, while these weather events are viewed as hazards, researchers argue that the winter season that spawns them is not. The same reasoning that they apply to winter, they apparently also apply to El Niño.

People have coped with seasonal changes for at least as long as societies have existed. So, they have already learned to expect the typical weather extremes that are likely to occur in a specific location during winter (cold temperatures, frost, snow, ice, sleet, etc.). Even though a specific winter may prove to be either longer or shorter, warmer or colder, earlier or later, milder or more severe than expected, winter gives rise to many hazards. Despite these uncertainties, societies and individuals have learned to put their coping strategies into action instinctively when winter approaches. They adjust their perceptions on a seasonal basis about how they expect their normal seasonally based activities to be altered. The more lead time available to respond to the impacts of winter, the better prepared societies are likely to be for most of the weather extremes that the winter climate might bring. While many people do not explicitly consider the seasons as natural hazards in and of themselves, they do so implicitly, because knowing a seasonal change is coming serves as an early warning to both society and individuals to prepare for a different set of possible seasonal, climate-related problems. Therefore, a strong argument can be made for considering winter as a hazard.

By analogy, using the notion of "winter as hazard spawner" as an example, one could make similar arguments for El Niño. Knowing that an

El Niño is coming provides a very early warning about possible changes in regional climate conditions and, therefore, about impacts on human activities and ecological processes that are likely to result. If an objective of managing the impacts of natural hazards is to reduce the adverse aspects of those impacts, then by viewing El Niño as a hazard (in the sense that it spawns hazards), societies can have an earlier start in determining how best to cope with those hazards when they occur. The same arguments that were made for El Niño can be made for La Niña (the ENSO cold event) as a hazard spawner.

El Niño and La Niña extend across several seasons, generating changes within the different seasons as they progress. Media articles on El Niño are in fact beginning to refer to an "El Niño winter," a winter that enhances the probability of "normal" wintertime hazards in some parts of the globe, while reducing the likelihood of such hazards elsewhere.

# 3 A tale of two histories

For the first two-thirds of the twentieth century, El Niño and the Southern Oscillation were studied and discussed as separate natural processes. Some researchers were interested in El Niño events in the eastern equatorial Pacific Ocean, while others focused on the Southern Oscillation, occurring in the atmosphere from the Indian to the Pacific Ocean. It was not until the mid-1960s that, building on the work of others (e.g., Hendrick Berlage, Sandy Troup), Jacob Bjerknes, an atmospheric scientist at the University of California at Los Angeles, recorded his ideas about linking the physical mechanisms of these two seemingly separate processes, one in the ocean and one in the atmosphere. To better understand El Niño in its broadest (that is, basin-wide) context, it would be useful to run through the history of the development of interest in El Niño and in the Southern Oscillation. The ***Appendix*** presents a time line in brief of specific key developments in our understanding of El Niño, La Niña, and the Southern Oscillation.

## History 1: Interest in El Niño

Interest in the impacts of the El Niño phenomenon, as such, goes back at least to the middle decades of the 1800s. At that time, its adverse effects on the so-called guano birds (i.e., seabirds such as cormorants, gannets, and pelicans) and on guano production (bird droppings used as a fertilizer for agriculture) had already been observed. Guano deposits that had built up over millennia were "mined" along the Peruvian coast throughout the second half of the nineteenth century, despite the decline in European demand for it. By 1900, Peruvian authorities were alarmed that this valuable export commodity (guano) was, in essence, being mined at an alarming and unsustainable rate; that is, it was being mined at a rate faster than it was being produced. As a result, in the first decade of the twentieth century, the Peruvian government established a Guano Administration Company to oversee the protection of the guano bird population and to control the mining of guano (Figures 3.1 and 3.2).

*Figure 3.1. Birds built a mountain of guano on the Central Chincha Islands over two and a half millennia; men carted it away in a few years. An orgy of exploitation in the 1860s stripped Chincha of its valuable cover. Measured against the workers, the heap is nearly 20 meters high. In some places the guano rose twice that much above bedrock; earlier digging reduced this pile. This old photograph shows Chinese laborers at work in ca. 1860. (Neg. No. 311830, photocopied by H. S. Rice, courtesy Department of Library Services, American Museum of Natural History.)*

Guano birds along the Peruvian coast live off of the fish populations that dwell near the ocean's surface (these are called pelagic fish), primarily the anchoveta. The cold oceanic conditions in Peru's coastal waters are usually optimal for anchoveta populations but are occasionally perturbed by El Niño. El Niño-related changes in physical, biological, and social conditions can be devastating for the anchoveta, and, in turn, to scabird populations. Some of these processes are described briefly in the following paragraphs.

*Figure 3.2. North Island (Chincha Islands) photo taken in ca. 1860, depicts a fleet in the foreground and guano mountain and stacks of guano on the horizon. (Neg. No. 311820, courtesy Department of Library Services, American Museum of Natural History.)*

*Physical setting*

The rotation of the Earth, combined with the winds that tend to blow toward the equator and offshore up the west coast of South America, pushes coastal surface water westward away from the continent and toward the central Pacific. As a result, cold water is drawn up from the ocean's depths to replace the warmer displaced surface water. This process is referred to as coastal upwelling. Coastal upwelling processes create regions in the ocean that are biologically highly productive oases (Figure 3.3).

The upwelling of deep cold ocean water brings a variety of chemicals into the sunlit layer near the ocean's surface. They are converted through photosynthesis to nutrients for phytoplankton which are at the bottom of the marine food chain. The plants are eaten by zooplankton and fish populations, which in turn are then consumed by guano birds. As one scientist noted in the first decade of this century, the anchoveta "are preyed

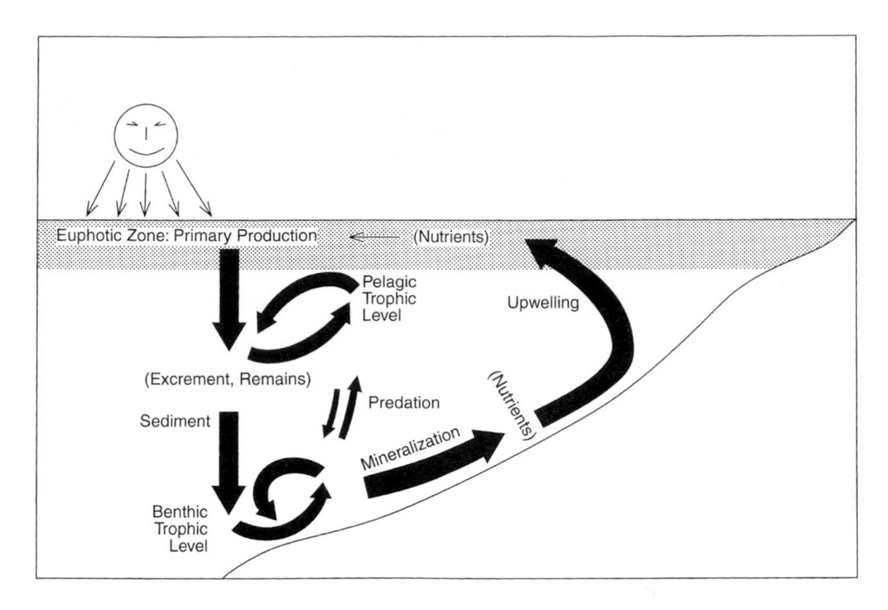


*Figure 3.3. Nutrient cycling in an idealized coastal upwelling system.*

upon by the flocks of cormorants, pelicans and gannets, and other abundant sea birds . . . The anchoveta, then, is not only . . . the food of the larger fishes, but, as the food of the birds, it is the source from which is derived each year probably a score of thousands of tons of high-grade bird guano" (Coker, 1908, p. 338).

The coastal upwelling phenomenon usually occurs along the western coasts of continents (with the exception of coastal upwelling along Somalia's coast in northeast Africa) in both the Northern and Southern Hemispheres. Coastal upwelling regions from around the globe make up about 0.1% of the ocean's surface area but provide around 40% of all the commercial fish captured globally. The cold water that upwells along the coast tends to suppress rain-producing processes in the atmosphere and, as a result, upwelling regions are usually found adjacent to coastal deserts. So, although coastal deserts appear to be harsh and barren environments from a societal perspective, the productivity that they did not get on land they found in their adjacent coastal waters. The location of major coastal upwelling regions are shown in (Figure 3.4).

### *Biological setting*

Biological productivity in the marine environment can be measured in terms of the rate of fixation of carbon by photosynthesis. The productivity of an upwelling ecosystem is measured in part by the amount

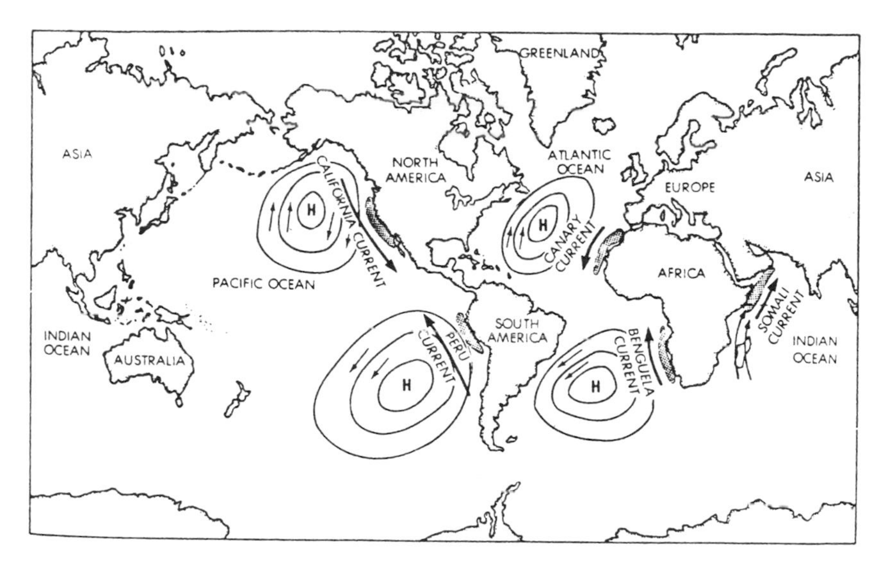


*Figure 3.4. Major coastal upwelling regions of the world and the sea level atmospheric pressure systems that influence them. (After Thompson, 1977.)*

of nutrients that is brought into the sunlit layer near the surface. According to David Cushing (1982, p. 19),

> Each [upwelling region] moves poleward as spring gives way to summer and each is two or three hundred kilometers broad in biological terms, even if the prominent physical processes are confined to a band within about 50 km of the coast.

Of these highly productive marine ecosystems, Peru's is considered to be one of the best not only in terms of the rate of fixation of carbon but of tonnage of fish caught as well (mostly anchoveta during non El Niño periods) (Figure 3.5). Before 1960, Peru was not noted for its commercial fishing activities, even though its commercial fish catches had doubled each year in the 1950s. However, from the mid-1960s to the first couple of years of the 1970s, it had become the world's number one fishing nation.

When El Niño events occur, coastal upwelling processes in the eastern Pacific are altered to such an extent that behavior within and among species becomes modified in major ways. Anchoveta, for example, disperse and migrate as well as dwell deeper in the ocean. Patterns of reproduction and migration change for the various fish species, with some reproducing less in the temporarily altered marine environment. Warm water species temporarily invade the waters of the eastern equatorial Pacific. Some fish populations such as sardines fare well in the new but temporary warm

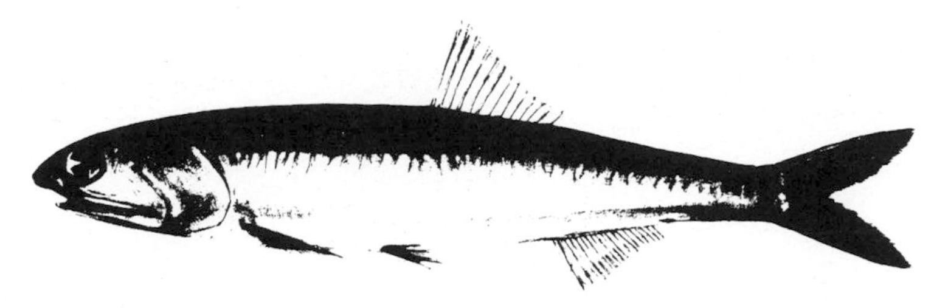

*Figure 3.5. Anchoveta (*Engraulis ringens *Jenyns). Adult, actual size 17 cm. (Art courtesy of the Instituto del Mar del Peru.)*

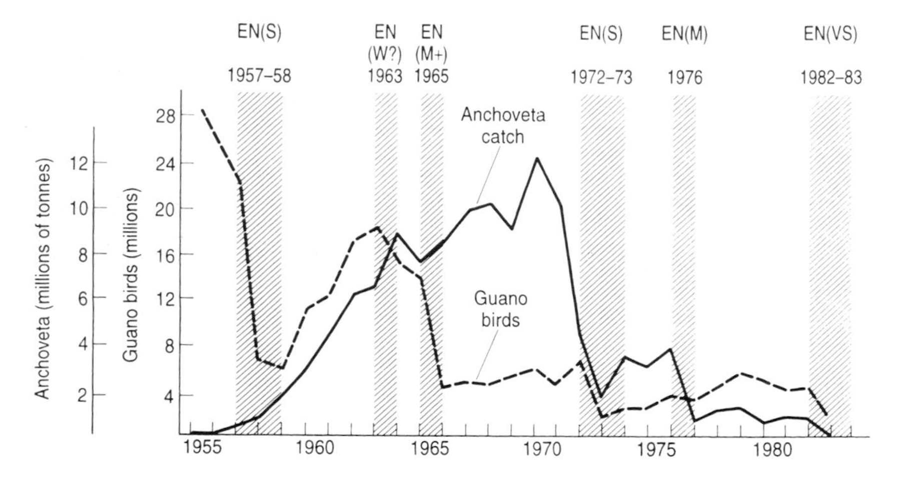


*Figure 3.6. Fluctuation of the anchoveta catch and guano bird populations (cormorant, gannet, and brown pelican) off the coast of Peru in relation to El Niño (EN) events. The letters signify the strength of the event: W, weak; M, moderate; S, strong; VS, very strong. (From Jordán, 1991, reprinted with permission of Cambridge University Press.)*

surface water environment. More specific to Peruvian interests, the standing stock of anchoveta (that is, the total population from which future generations are to be produced) becomes reduced for a variety of reasons, including higher mortality and lower fertility.

As a result of changes in the behavior of the anchoveta, the fish population becomes much less accessible to the guano birds, which causes the starvation and death of hundreds of thousands to millions of birds, depending on the magnitude and intensity of the particular El Niño episode. Figure 3.6 provides an example of the adverse impacts on Peruvian guano birds of the combination of El Niño events and heavy commercial fishing pressures over the span of a few decades.

*Figure 3.7. Modern Peruvian fishmeal-processing plant situated along the barren Peruvian coast about two hours south of Lima. (Photo M. H. Glantz.)*

*Societal setting*

The Guano Administration Company, created in 1909 by the Peruvian government to manage the guano resource, and its main supporters among the Peruvian agricultural elite managed for many decades to block the development of large-scale commercial anchoveta fishing ventures. Apparently, they were able to argue successfully within the nation's highest political circles that there were not enough fish in Peru's coastal waters to sustain both a viable guano-mining industry *and* a viable anchoveta fishing sector. These predators – birds and fishermen – would be competing for the same resources in order to "survive". It is important to note that the Peruvian anchoveta captured by the fishing boats were not caught in order to be consumed directly by humans. They were caught to be processed into fishmeal (Figure 3.7) for use as an animal feed supplement for export primarily to the rapidly expanding North American poultry industry. The anchoveta was also a source of fish oil for Peru's domestic market.

In the early 1950s, however, the entrepreneurs interested in developing a Peruvian commercial fishery finally convinced politicians at the time to allow them to establish a commercial fishing industry, winning out over those who opposed its development. The arguments of Peruvian investors in favor of establishing a national fishery received a major boost when the California sardine fishery collapsed from overfishing. The adverse impacts on a California fishing community were captured in the 1940s and '50s in American writer John Steinbeck's novels *Cannery Row* (1945) and *Sweet Thursday* (1954). This collapse sparked an increase in the demand by the USA for the need to import anchoveta fishmeal. At the same time the collapse also made available for sale California's idled (and now useless) fishmeal-processing plants and fishing vessels. The lure of being able to sell fishmeal for sorely needed hard currency (i.e., dollars) in the international marketplace put great pressure on Peruvian policymakers to agree to open up Peru's coastal waters for exploitation by a commercial fishing fleet. In fact, it has been rumored that in 1953, the first fishmeal-processing factory had been brought into Peru from a defunct California sardine fishery. According to those rumors, it was clandestinely reassembled at a remote site along Peru's arid coast without government permission. Figure 3.8 provides an overview of the Peruvian anchoveta fishery.

Robert Cushman Murphy, a zoologist with the American Museum of Natural History in New York City, studied guano birds along the Peruvian coast for several decades. In a special report prepared in 1954 for the Guano Administration Company for presentation to the Peruvian government, he noted that the development of a large-scale, commercial fishery based on the exploitation of the anchoveta population would, *without a doubt*, lead to the demise of Peru's guano bird population. He based his argument on the view that guano birds consumed only the number of fish that they needed to satisfy their immediate food needs, whereas fishermen in their gas-fueled boats were insatiable predators. Fishermen would take from the ocean as many fish as their nets could capture and their boats could hold, and that they could sell to "insatiable" processing factories (Murphy, 1954). From the official opening of the anchoveta fishery in the early 1950s, a doubling of the officially reported anchoveta catch was repeated every year over the preceding year from the mid to late 1950s, as noted earlier. However, it is very likely that many landings were not reported, so the actual impact of the nascent Peruvian fishing sector on the anchoveta population is difficult, if not impossible, to determine.

Although a strong El Niño took place in 1957–58, there was little, if any, direct reference to it in the popular press outside of Peru's borders. The most obvious reason is that there were no visible adverse effects on the productivity of the Peruvian anchoveta fishery. No news was good news. Apparently, by the late 1950s the anchoveta fishery had not as yet grown very large compared with the available supply, although it was rapidly expanding at the time. Thus fish landings then were still well below what scientists considered to be the fish population's maximum sustainable yield (MSY). MSY refers to the yield of fish that could be sustained over an

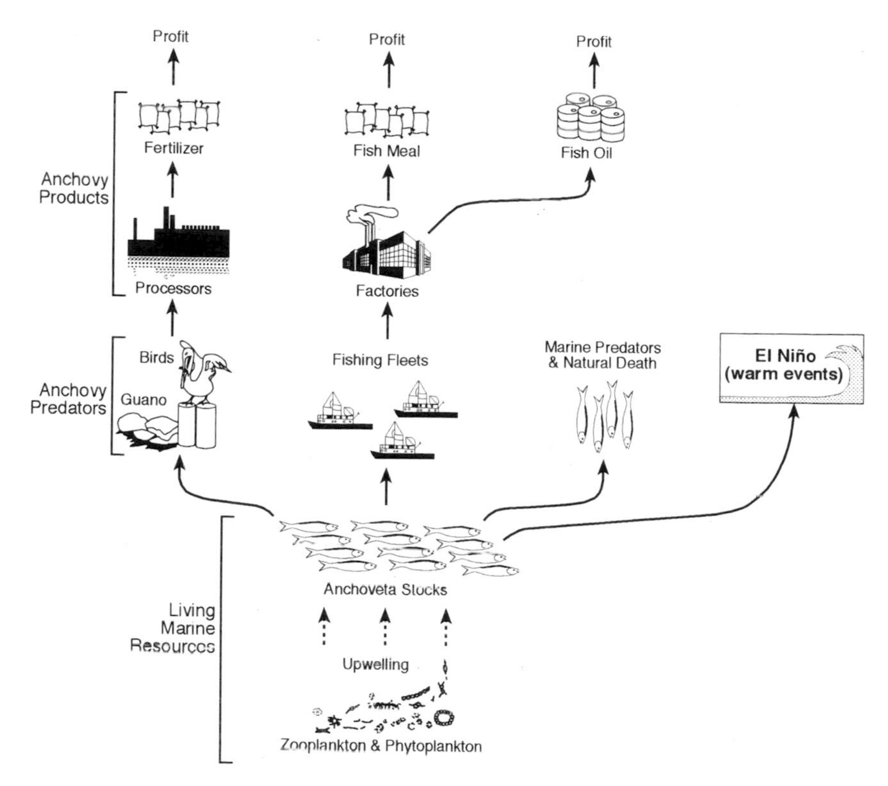


*Figure 3.8. A diagram showing biological, economic, and social components involved in human exploitation of the natural fertility of the Peru Current as of the early 1970s. (Based on Paulik, 1981 in* Resource Management and Environmental Uncertainty: Lessons from Coastal Upwelling Fisheries, *ed. M. H. Glantz and J. D. Thompson. Copyright 1981 John Wiley & Sons, Inc. Reprinted by permission of John Wiley & Sons, Inc.)*

indefinite period of time without destroying the fish population. Today, researchers would refer to this as sustainable development. Hence entrepreneurs (Peruvian investors from all walks of life) and fishermen in the newly emerging Peruvian fishing industry showed little concern for or, more correctly, awareness of, El Niño or its potential impacts on the fish population that they were hell-bent on exploiting.

As a result of the 1957–58 event and its effects on the upwelling process, the usage of the term El Niño was deliberately broadened in its oceanographic context by the scientific community. It was viewed as a phenomenon common in one form or another to other major coastal upwelling regions, and especially to the one off the coast of the US Pacific Northwest states of California, Oregon, and Washington. As early as 1959, oceanographer Warren Wooster commented on the concern that had been raised at that time about the use and misuse of the term El Niño. According to Wooster,

> Some readers . . . have objected to the use of the name El Niño to identify the general phenomenon, feeling that previous usage restricts the term to the Peruvian Coast. If a more appropriate generic term can be found, I would recommend its use. I have used El Niño in a broad sense to emphasize that the Peruvian Niño is not a unique phenomenon, but is rather merely a striking example of a wide-spread occurrence. (Wooster, 1959, p. 45)

By the mid-1960s, marine biologists and foreign fishery consultants at the then recently established Peruvian Marine Institute (IMARPE) in Callao had become concerned about the expanding and uncontrolled levels of fishing. Pressures to keep open access (free-for-all) to the anchoveta resource came from fishing-boat owners, Peruvian banks that had loaned funds for boats and equipment, from fishermen, and from the highly competitive Peruvian fishmeal factories. Biologists also realized that El Niño events, combined with growing fishing pressure, could increase the vulnerability (and weaken the resilience or ability to recover) of the basic population of fish from which future populations were supposed to come.

While the 1965 El Niño event appeared to have reduced anchoveta landings only slightly, it had a devastating impact on the guano bird population from which it has yet to recover (see Figure 3.6). This event was apparently the first "wake-up call" to some members of the fishing industry and the Peruvian government. It alerted them to potential fisheries problems of which their national marine biologists were already aware.

From the perspective of the anchoveta, El Niño events are like other predators – guano birds, fish predators, fishermen and fishmeal factories. El Niño events, too, take their share of anchoveta. With fishing pressures on the anchoveta mounting, policymakers and fishery managers were coming to realize (ever so slowly) that this fish population was not a limitless resource.

In response to those emerging concerns, biologists calculated at that time – the mid 1960s – that the maximum sustainable yield of the anchoveta population was about 9.5 million tonnes per year. They calculated that the guano birds consumed about 2 million tonnes, with the other 7.5 million tonnes going to the fishermen. At least one American fisheries scientist, as well as some Peruvians, actually called for the deliberate destruction of the remaining guano bird population in order to "free up" the 2 million additional tonnes of anchoveta for Peruvian fishmeal-processing plants. Although this suggestion may have been seriously considered by some decisionmakers, it was not acted upon.

In 1968, the Peruvian military overthrew the government. Up to the time of this *coup d'état*, the fishing sector had been under the administrative jurisdiction of the Ministry of Agriculture (Hammergren, 1981). As the Ministry's name implies, most of its attention was focused on the agricultural sector, thereby downplaying the importance of the fishing sector's management needs. In its heyday in the late 1960s, the fishing sector generated almost one-third of Peru's foreign exchange earnings. Hard currency earnings (the US dollar, British pound, French franc, or German mark) are extremely valuable to developing nations, because they enable countries to import new technologies and other foreign goods. To many Peruvians, the number of anchoveta in their waters was infinite. Successive Peruvian governments up to the time of the 1968 coup appeared to have had little awareness of just how important the fishing sector was to Peru's economy.

From low-level Peruvian bureaucrats to high-level policymakers, all of whom were unfamiliar with either fish biology or the already-known consequences of the overexploitation of a fish population in other parts of the globe, the coastal ocean was perceived to be an endless source of fish and, therefore, of foreign exchange. Perceptions such as these caused Peruvian policymakers to disregard the advice of their own national scientists in the late 1960s about the urgent need to tighten up the management of the fishery in order to save the resource itself. Peruvian scientists warned that the failure to better manage their living marine resources would probably lead to the collapse of the fishing sector, as had happened throughout the twentieth century in many other commercial fisheries worldwide.

Owing to a misbelief that "experts come from out of town," the Peruvian government felt compelled to seek the advice of foreign scientists by convening panels of international fisheries experts to evaluate the scientific advice that policymakers were receiving from Peruvian scientists. Time after time, the various panels of foreign fisheries experts issued reports that upheld the research findings and recommendations of their Peruvian colleagues: too many boats and too many factories were dependent on the increasing exploitation of a dwindling number of fish. By the time the Peruvian government finally got the message that it should have believed its own scientists, the anchoveta fishery was well on its way to collapse.

## 1972 and after

In my view, the year 1972 should be called "the year of climate anomalies". For various reasons a set of adverse climate-related events worldwide captured news headlines. Those events had deleterious effects on global food production and global food security. Collectively, these anomalies had a major impact on policymakers' perceptions about the declining ability of various countries around the world to feed their citizens (Garcia, 1981). The former Soviet Union, for example, registered one of its worst shortfalls in grain production, as a result of severe drought. It resorted to major grain imports from the USA, especially wheat and corn, which in turn exacerbated the scarcity of these commodities on the world market (Trager, 1975). Droughts also occurred that year in Central America, the Sahelian zone of West Africa, India, the People's Republic of China, and in parts of Australia and Kenya. As a result of the 1972 climate anomalies, along with other socioeconomic and political factors, global food production per capita and global food reserves had declined for the first time in more than 20 years (Brown and Eckholm, 1974).

In addition to these events and to the decline in Peruvian fish landings, there was a simultaneous decline in fish catches in other parts of the world. Fisheries biologists and national policymakers were forced to rethink their previously held assumption that the oceans would become a major source of food that could supplement food production on land (see, for example, *Mosaic*, 1975; Thompson, 1977).

In retrospect, the 1972–73 El Niño had made a bad global food supply situation even worse. Its impacts on biological productivity in Peru's coastal upwelling ecosystem, combined with the decimation of the anchoveta population because of heavy fishing pressure, contributed in a major way to the collapse of Peru's fishing industry. Because of this El Niño, fishmeal was available only in reduced quantities and at very high prices in the marketplace. The second preferred choice as an animal feed supplement of the poultry industry, a major consumer of fishmeal, was soymeal. To meet the sharp increase in market demands for feed supplements created by the high price of fishmeal, US farmers planted soybeans instead of wheat. This was done at a time when a major global food crisis was emerging and, as a result, wheat was in great demand. However, the price that farmers received for soybeans outpaced the price for wheat. The implication of this shift in production from wheat to soybean was serious for the global food situation. At a time when the demand for humanitarian assistance from several drought-plagued developing countries had sharply increased, farmers in North America were growing crops as feed supplement for animals instead of producing grains for human consumption.

Numerous devastating droughts and other weather, climate, and climate-related anomalies worldwide accompanied the 1972–73 El Niño (Figure 3.9). Rightly or wrongly, many of these were associated by different people with that El Niño. These particular impacts prompted some countries to develop an interest in, and respect for, an improved understanding of El Niño events and their direct and indirect consequences.

Two weak El Niño events occurred along the Peruvian coast later in the

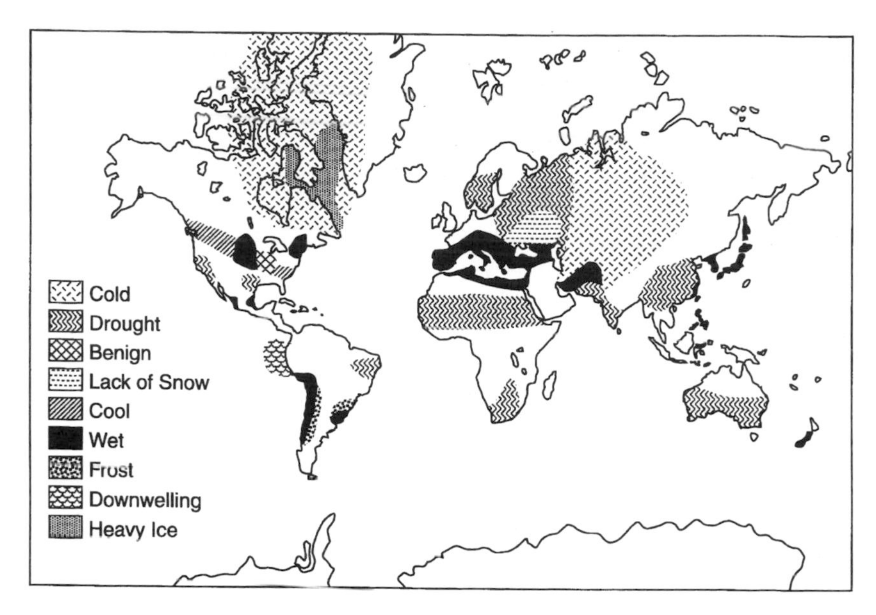


*Figure 3.9. Global climate anomalies map for the year 1972, one of the first global composite maps. This was a year of climate anomalies and of a major El Niño event with substantial impacts on Peruvian fisheries. (After McKay and Allsopp, 1976.)*

1970s. These were followed in the early 1980s by what was then called the "El Niño of the century". The 1982–83 El Niño sparked major interest once again in the interactions between oceanic and atmospheric processes in the equatorial Pacific Ocean and the linkages of those processes to climate anomalies and climate impacts on societies around the globe. That attention has since been heightened by subsequent moderate El Niño events, one in 1986–87, another that began in 1991–92 and by a second "El Niño of the century" in 1997–98. I would argue that the 1982–83 event was the "El Niño of the scientists", because it sparked great interest among scientists in the phenomenon. I would then argue that the 1997–98 event was the "El Niño of users," because it led to increased demands by decisionmakers in a wide range of economic sectors of society for El Niño-related forecasts all around the globe.

Even today, almost two decades after the extraordinary 1982–83 event, scientists and policymakers are still seeking to generate interest in El Niño research and impacts, respectively, and, as a result, they continue to refer to the 1982–83 event as the proverbial "yardstick" against which other subsequent events of the twentieth century have been measured. The intense 1997–98 El Niño is likely to become the new yardstick, at least for the first few decades, of the twenty-first century.

## History 2: Interest in the Southern Oscillation

Peruvian geographer Victor Eguiguren (1895) suggested that there were linkages between excessively heavy rains and flooding in the northern coastal city of Piura and the occasionally recurring warm coastal current then called El Niño. Around the same time, the last decades of the nineteenth century, observers of variable weather conditions in India and Australia were generating hypotheses about how regional changes in the atmosphere contributed to recurrent famines in India and to droughts in northern and eastern Australia. They focused on how variations in sea level pressure or changes in the land's surfaces such as changes in the extent of snow that covered the Eurasian land mass (centered on the Himalayas) might relate to failure of the monsoonal rains in India.

In the mid-1800s, the British Empire encircled the globe and the moral, if not political, responsibility for food security in its colonies fell on Britain. As a result of the major drought- related devastation wrought by famine in India in 1877, some researchers focused on trying to identify ways to forecast climate anomalies (e.g., droughts and floods) mainly for the purpose of forecasting food production prospects for a season or from one year to the next year.

In 1884 Henry Blanford, the main meteorological reporter for the Government of India, reported to the Indian Famine Commission that he could not find a way to use solar activity to predict weather in India. However, he believed that the increases in snow cover in the Himalayas and on the Eurasian land mass in springtime were responsible for the failure of the monsoonal rains on the Indian subcontinent. Research continues today on the impact of the areal extent of Eurasian snow cover on global climate variations from year to year.

Charles Todd, a government astronomer in South Australia noted in the late 1880s that droughts in India (often referred to as "the failure of the monsoons") seemed to occur at the same time as droughts in northern and eastern Australia. Todd suggested the possibility of connections between the two through the atmosphere (Todd, 1893).

Norman Lockyer, a British astronomer and meteorologist, also active in the last few decades of the nineteenth century, searched for linkages between solar activity (sunspots) and Indian rainfall. He tried to correlate Indian rainfall with rainfall failures in various parts of Australia, because he, too, had noticed that droughts in northern Australia seemed to occur when there were droughts on the Indian subcontinent (Lockyer and Lockyer, 1904).

At the end of the nineteenth century, mathematician Gilbert Walker became the Director- General of Observatories in India. When he retired in 1924, he returned to England and focused his attention on attempts to

> The British government began to pay considerable attention to Indian monsoon variations after the devastating drought of 1877, which we now know was an El Niño year. Another major drought over India occurred in 1918, also an El Niño year, while Walker was the head of the India Meteorological Department. While looking for teleconnections with Indian monsoon rainfall, and not being aware of El Niño, Walker discovered the Southern Oscillation. The relationship between El Niño and monsoons still remains elusive, especially because the major Indian droughts occur six months before the peak phase of El Niño. the quest for predicting monsoons has been further complicated by the fact that the El Niño – monsoon relationship has qualitatively changed [since the late 1970s].
>
> J. Shukla, Institute of Global Environment and Society, 1996

predict the behavior of the Indian monsoon and, later, on changes in sea level pressure patterns across the Indian and Pacific oceans. The particular pattern he observed, which in 1924 he labeled "the Southern Oscillation," was the result of a seesaw-like oscillation of sea level pressure changes at various locations across the tropical Pacific basin. He used pressure records from locations such as Darwin (Australia), Canton Island in the equatorial central Pacific, and Santiago (Chile). Today, researchers have chosen to use the difference between sea level pressure at Tahiti (French Polynesia) and at Darwin (Figure 3.10). The quantitative differences between the centers of these distant pressure systems (i.e., Tahiti pressure minus Darwin pressure) have been converted into an index now called the Southern Oscillation Index (SOI). Usually, there is a low pressure system in the region of Indonesia and northern Australia, centered near Darwin. This system brings storminess to the region, providing some parts of Australia, the driest inhabited continent on Earth, with the moisture they sorely need for agricultural production and livestock management for sustaining their settlements, wildlife, and ecosystems. At the same time, there is usually a high pressure system in the southeastern Pacific, centered near Tahiti. The sea level pressure at each of these two points is related to the other. When the sea level pressure at Darwin is low, usually (but not always) the pressure at Tahiti is high, and vice versa. El Niño is related to the negative phase of the SOI and La Niña to its positive phase. Therefore, the difference of sea level pressure (Tahiti minus Darwin) is used as an index that characterizes the ENSO cycle of warm and cold events (Figure 3.11).

The SOI has reliably been associated with a number of climate-related events: when the sea level pressure at Darwin increases, the likelihood of drought in the Australian–Indonesian region increases; when the pressure at Tahiti decreases, the likelihood of more rainfall in the equatorial central

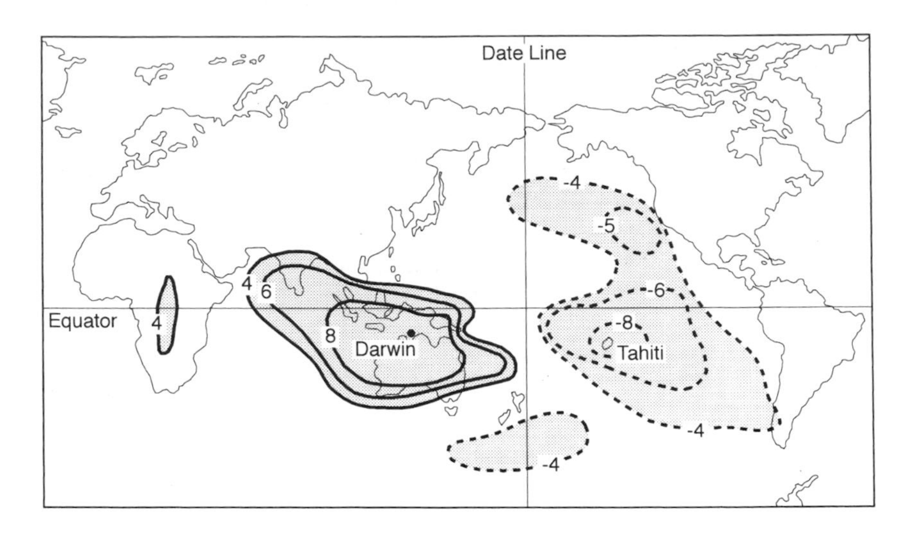


*Figure 3.10. Tahiti and Darwin are at opposite ends of the Southern Oscillation's seesaw, and so the difference in pressure between them is used to measure the Southern Oscillation. The numbers represent a statistical measure called the correlation coefficient. The figure shows that the pressure variation at Tahiti is as closely related to the variation at Darwin, as are locations near to Darwin but with the opposite sign (i.e., if the pressure is high at Darwin, it is low at Tahiti and vice versa). (After Rasmusson, 1984a.)*

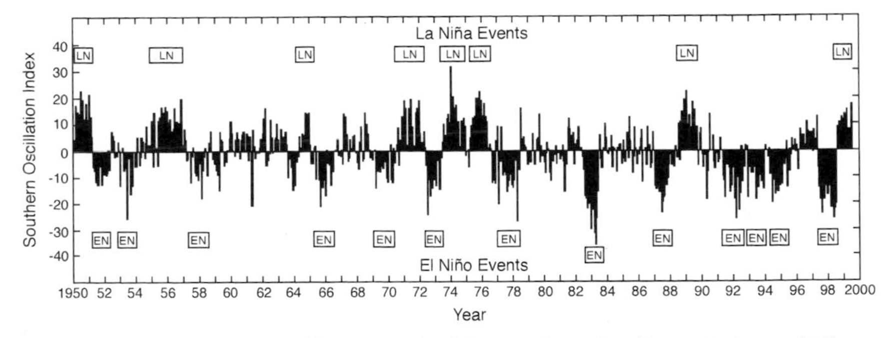


*Figure 3.11. Monthly averages of the Southern Oscillation Index and El Niño (warm) events and La Niña (cold) events, 1970–90. El Niño and La Niña years are identified with horizontal bars. (Adapted from Nicholls, 1993.)*

Pacific region increases. These particular sea level pressure changes appear to set the stage for the possible (but not certain) onset of El Niño. Pressure changes are accompanied by changes in wind speed and direction, shifts in the location of pools of warm and cold ocean surface water, changes in the strength of equatorial and coastal upwelling, and shifts in the location of biological productivity in the ocean that, in turn, alter the locations of various commercially exploited fish populations.

More than 80 years ago, Gilbert Walker used statistical methods to identify weather anomalies linked to the Southern Oscillation. Following on the heels of others, Walker identified many apparent associations or relationships (called correlations) between changing atmospheric pressure patterns at sea level in the Australasian region and rainfall patterns in such distant locations as the Indian subcontinent, Africa, and South America. Because of his strong mathematical background, he was able to use those methods appropriately, and he achieved a high standard that has not necessarily been matched by many subsequent researchers.

For the most part, Walker's correlations have survived numerous re-evaluations and challenges by succeeding generations of researchers using a variety of modern statistical measures and computer technologies. Decades later, to honor Sir Gilbert Walker, Jacob Bjerknes named an important atmospheric circulation pattern in the Pacific, the Walker Circulation, which links the Southern Oscillation with sea surface temperatures.

In Australia, the Southern Oscillation has been monitored closely as a tool for forecasting the likelihood of rainfall in various parts of Australia several months in advance, regardless of extreme warm or cold events in the Pacific. One researcher (see Bacastow *et al.*, 1980, p. 67) noted that "fully developed El Niño events such as occurred in 1965, 1969, 1972, and 1976 are observed to coincide with a minimum of SOI" (Figure 3.12). He also suggested that there was support for the hypothesis that "all minima of the SOI are accompanied by El Niño-type conditions, even if a fully developed El Niño does not occur". The major point is that, although there is not a perfect one-to-one correlation between observed El Niño events and the SOI, this relationship is *quite strong*. Despite the long SOI record going back to the mid-1870s, most recent researchers and forecasters have been moving away from a reliance on it as the primary indicator of the onset of an El Niño or a La Niña. They have chosen, instead, to rely on the monitoring of changes in ocean temperatures around the globe. One persistent argument against an overreliance on the SOI has been that Tahiti is not truly representative of pressure changes in the eastern equatorial Pacific Ocean. Another argument against a heavy reliance on the SOI is the fact that teleconnections between the Pacific and climate anomalies in distant locations can be directly affected by changes in sea surface

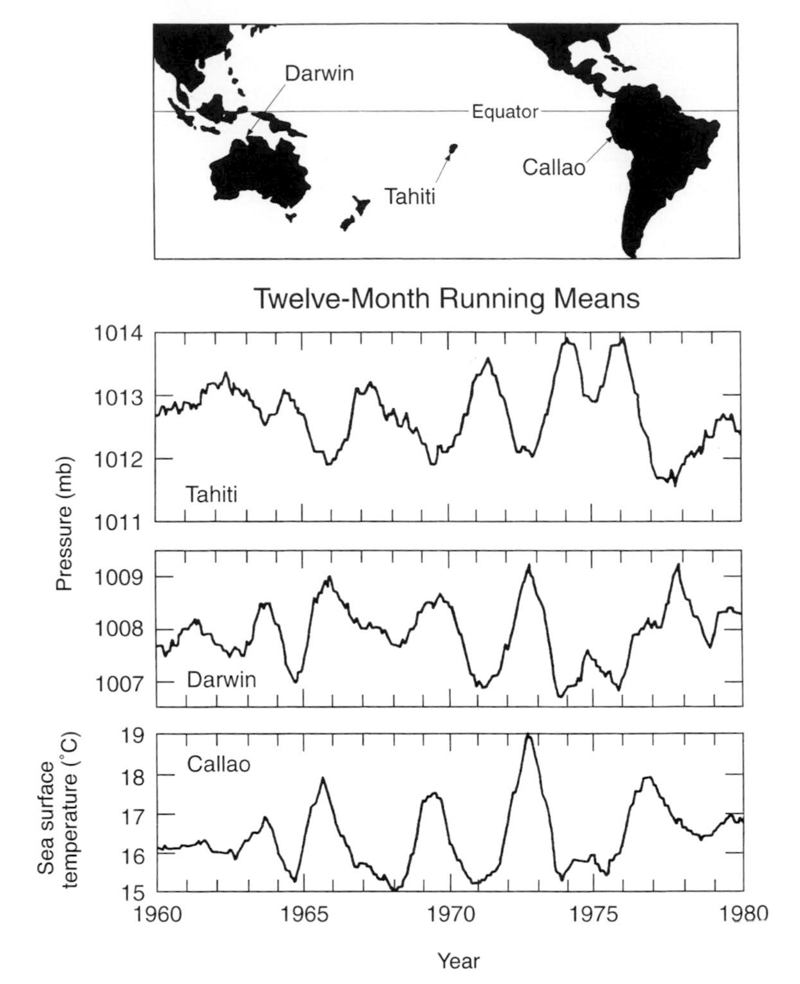


*Figure 3.12. Smoothed curves showing changes in atmospheric pressure at Tahiti and Darwin and changes in sea surface temperatures in the eastern equatorial Pacific Ocean at La Punta/Callao, Peru. Callao is the site of IMARPE, the Peruvian Marine Institute. (After Hansen, 1990.)*

temperatures in other oceans. Problems in forecasting the onset and severity of the 1997–98 El Niño has prompted some Australian research groups to move away from a heavy reliance on the SOI for forecasting droughts in Australia.

*Teleconnections*

The word "teleconnection" first appeared in the scientific literature in the mid-1930s and was used to describe physical relationships affecting climate conditions in the North Atlantic region (Ångström, 1935). To be sure, the notion behind it – that weather changes at one location might be related to weather changes at other remote locations – had existed for a long time. As noted earlier, people have been fascinated since at least the latter part of the nineteenth century by the prospect of identifying linkages among weather events in various parts of the globe. At first, qualitative assessments and scientific hunches were relied upon to identify them, as opposed to the sophisticated quantitative methods that are being used today.

Since the onset of the twentieth century, the identification of teleconnections has become a subfield of scientific research as well as a pastime for non-weather specialists and the media in their attempts to forecast regional and local weather and climate anomalies some months, seasons or years in advance. Although Sir Gilbert Walker is perhaps the best known of those early researchers who sought to identify such linkages among temperature, rainfall, and pressure anomalies at some distance from each other, we now know he was by no means alone.

Walker's methods and findings were challenged, if not ignored, by many of his contemporaries right up to the time he died and even shortly thereafter. The way that many in meteorology looked at Walker's teleconnections work as late as 1959 (Brown and Katz, 1991) was succinctly captured in an excerpt from his obituary that appeared in a 1959 issue of the *Quarterly Journal of the Royal Meteorological Society* (**85**, 186):

> Walker's hope was presumably not only to unearth relations useful for forecasting, but to discover sufficient and sufficiently important relations to provide a productive starting point for a theory of world weather. It hardly seems to be working out like that.

It was only a decade or so later that Walker's contributions to teleconnections research overshadowed the comments of his critics. Today, the desire to identify teleconnections with a high degree of reliability is a key driving force behind increased interest in an improved understanding of El Niño and La Niña processes. Such an improvement would greatly enhance attempts at producing seasonal, interannual, and interdecadal climate and climate-related forecasts (e.g., forecasts related to food production, energy consumption, water use, severe storms, and infectious disease outbreaks). As one example, Colorado State University professor William Gray uses information on sea surface temperature changes in the equatorial Pacific to construct his hurricane season forecasts. He also relies on the "climatology" of El Niño for his projections. His statistical assessments of tropical cyclones in the Atlantic strongly suggest that, during El Niño years, fewer tropical storms and hurricanes can be expected. Following an El Niño (that is, in non-El Niño years), the number of tropical storms and hurricanes increases. This example notwithstanding, the search for teleconnections between extreme weather events at great distances from one another, as well as between such events and human welfare, has not been without its detractors. Even the reliability of some fairly robust El Niño-related teleconnections, such as droughts in northeast Brazil or mild winters in the northeastern USA, has been challenged. Nevertheless, research on teleconnections is clearly warranted because the potential payoff to societies of research activities that can identify reliable physical and social teleconnections far outweighs the costs associated with scientific programs to search for them.

## Concluding comments

Ever since the physical mechanisms that linked El Niño and Southern Oscillation phenomena were identified in the late 1960s, their futures had become intertwined. However, the apparent 5-year El Niño (i.e., 1991–95) raised new questions about just how entwined these two processes really are. It challenged the tightness of the linkage between El Niño in its traditional sense as a local sea surface warming off the Peruvian coast and the Southern Oscillation. Although El Niño captivates the attention of Peruvians and the Southern Oscillation is of major interest to Australians, the Pacific basin-wide El Niño has captured the interest and attention of researchers, policymakers and much of the public around the world. Keep in mind, however, that a community of researchers has focused its attention on the El Niño phenomenon only for the past 30 years or so. The use of historical and other proxy information about El Niño events enabled researchers to identify past occurrences. It was in the late 1990s that the scientific community had the expanded capability (and sustained interest) to monitor most of El Niño's field of action – the Tropical Pacific – more closely than ever before.

# Section II
# The life and times of El Niño and La Niña

# 4 The biography of El Niño

The use of analogies in carrying out scientific research is much more widespread than even scientists themselves might realize. To plan and run their experiments, they use such analogies as "slab ocean", swamp models, water sloshing around in a bathtub, "water-capturing buckets," and even the "greenhouse notion". On the real-world side, we often use analogies to make decisions. Most recently, Hurricane Floyd in the tropical Atlantic was identified as being analogous to (i.e., having the same strength as) Hurricane Andrew 7 years earlier but as being three times its size spatially. The analogy of worse damage than that wrought by Andrew sparked widespread compliance by citizens with respect to evacuation along Hurricane Floyd's path. This particular analogy worked to motivate the public.

Analogies can also be used to provide a reader with a general idea of how the ocean and the atmosphere interact in the tropical Pacific Ocean along the equator. Analogies can provide a nice overview but they generally lack enough detail to provide an accurate picture of what is really going on in the region. For example, some media writers who popularize discussions of oceanic processes use a bathtub as an analogy for the Pacific Ocean. They note that a pool of warm water sloshes around the central Pacific from west (non-El Niño condition) to east (El Niño condition) and back again. For its part in this air–sea interaction, the atmosphere, heated by a warm pool of water, has been likened to a stove in a closed room.

Tribbia (1995) described the analogous process in the tropical Pacific with respect to the circulation of the atmosphere. This analogy has been used by scientists for at least the past 100 years (Figure 4.1).

> The stove in the corner of a room is heating up a portion of air in that room; the warmed air is lighter and rises to the ceiling of the room, crossing laterally toward the window where it cools. It then sinks, reaches the floor, and is sucked back toward the stove.
>
> (Tribbia, 1995, p. 18)

The tropical atmospheric circulation operates in a similar fashion, if the

*Figure 4.1. Artist's rendition of atmospheric circulation in a closed room as idealized circulation of air across the Pacific at the equator. (From Tarr and McMurry, 1904.)*

stove (i.e., source of heat) is in the west end of the Pacific basin. When the ocean's surface is warmed in the western part of the tropical Pacific, "the heating of the atmosphere drives a cloud system (called a convective cell). Low-level air flows across at the base (from right to left), is heated within the cell and rises and flows out (from left to right) at high levels" (Gill and Rasmusson, 1983, p. 231; Ghil and Childress, 1987). To simulate El Niño, however, the stove would have to be moved out to the middle of the room. In this situation there would be rising motion in the middle of the room and sinking motion at both ends. Pressure would be lower in those parts of the room where warm air is rising. In the El Niño-like room, there is low pressure in the center and higher pressure at the sides. Winds flow from high pressure to low pressure.

## The Walker Circulation

Scientists consider "normal" conditions (or, more broadly, non-El Niño) to exist when the following condition of air–sea interactions prevails in the equatorial Pacific. In the western part of the Pacific basin off of the Philippines and Indonesia near the equator, there is a very warm pool of sea water at the ocean's surface, the warmest ocean water on the planet. This is specifically labeled by scientists as "the warm pool". It covers an area about the size of the continental USA. That pool extends downward from the surface to a depth of a couple of hundred meters to the zone in the ocean where there is a sharp temperature contrast between the warm waters above and the cold waters below. This zone of a sharp temperature change is called the thermocline. The sea level in the western Pacific is higher by several tens of centimeters than it is at the eastern edge of the basin. This is because of the strong trade winds that flow toward the west at the ocean's surface. The trade winds tend to "move" (some say push) water toward the western edge of the Pacific basin, where it piles up and is heated by the sun.

The large pool of warm water in the western equatorial part of the Pacific Ocean is a major source of heat that warms the atmosphere above it. This warming causes air to rise, which, by generating convection, produces rain-bearing clouds. As the warmed air rises to even higher levels of the atmosphere, it cools. The pressure difference between the west Pacific and the east Pacific moves the now-cooler air to higher altitudes (e.g., the tropopause) and pushes it toward the eastern part of the Pacific basin. The cool, dry air aloft ultimately descends in the eastern equatorial Pacific (what goes up must come down!). The rule of thumb is that the descending motion of the atmosphere (which is called subsidence) tends to suppress the conditions that could bring about cloud formation and, therefore, reduces the probability for rainfall. The descending dry air then moves westward near the Earth's surface, as a result of westward flowing wind action. It becomes warmed again by the ocean's surface, from which it is then able to pick up moisture and the warmed moist air begins to rise again in the western Pacific. This is called the Walker Circulation.

Meanwhile, under non-El Niño conditions (i.e., normal and La Niña), the thermocline is relatively close to the surface in the eastern part of the equatorial Pacific, actually surfacing in the Southern Hemisphere in summer and autumn, whereas in the western Pacific it is at a depth of about 200 meters (see Figure 5.4a, p. 70). Recall that the sea level in the east is normally several tens of centimeters lower than that in the western part of the basin. Upwelling of deep cold nutrient-rich water is strong along the coasts of Ecuador, Peru, and northern Chile, making sea surface temperatures in the eastern Pacific considerably colder than those in the west. This tends to reinforce the atmospheric mechanisms, inhibiting cloud formation and rainfall in the region. The strength of the winds in the equatorial region is heavily influenced by the seesaw-like sea level pressure differences across the equatorial Pacific, from Darwin to Tahiti.

## El Niño and the Walker Circulation

During an El Niño event, the Walker Circulation becomes modified in a major way. The westward-flowing surface winds across the equatorial Pacific basin weaken and in the western part of the basin they reverse and flow eastward. This enables water in the warm pool in the west to spread eastward. As the warm water shifts eastward, the sea level in the west begins to drop, and the sea level in the east increases. With the slowing down of the westward winds, the surface waters of the central and eastern Pacific become warmer. As this occurs, the thermocline also begins to shift, moving upward toward the ocean's surface in the west and deepening in the central and eastern equatorial Pacific. As the thermocline moves downward along the Peruvian coast, upwelling continues but the water brought up to the surface is warmer and less rich in nutrients (see Figure 5.4b, p. 70).

Meanwhile, the water in the western equatorial Pacific becomes a few degrees cooler, as the surface and subsurface waters in the central and eastern Pacific warm up. Because convective activity (cloud formation) follows the sea's warm surface water, clouds increase in the central and eastern Pacific, while they decline in the west. This displacement in convective activity generates droughts in Australia, Papua New Guinea and Indonesia, typhoons in the central Pacific, and heavy rains along the normally arid coast of northern Peru. These conditions can last from 12 to 18 months, until the westward flowing surface winds once again begin to strengthen, causing warm water to flow back toward the region of the western Pacific warm pool. The sea levels at both ends of the basin begin to change direction now rising in the west and falling in the east, as does the depth of the very important but out-of-sight thermocline. Strong upwelling returns to the equator and to the eastern Pacific boundary of coastal Peru.

Figure 4.2a,b summarizes the west-to-east interaction between the atmosphere and the ocean in the equatorial Pacific under "normal" (non-El Niño) conditions, and under El Niño conditions. The precise timing of the beginning of any particular El Niño event may not be well known, although there are several hypotheses about how to detect it. Once started, however, the processes that keep El Niño going, as well as the processes that end it, appear to be better known. In fact, once an El Niño has started, scientists usually have a better idea some months in advance about its potential impacts on some ecosystems and societies around the globe. This was the case for the 1997–98 event. Even though forecasters missed forecasting its onset, they were relatively more successful in identifying some of its worldwide impacts.

## The El Niño process over time

### *The phases of El Niño*

Early research on El Niño up to mid-1982 seemed to suggest that there was only one kind of El Niño, referred to as the "canonical El Niño". It started along the west coast of equatorial South America when sea

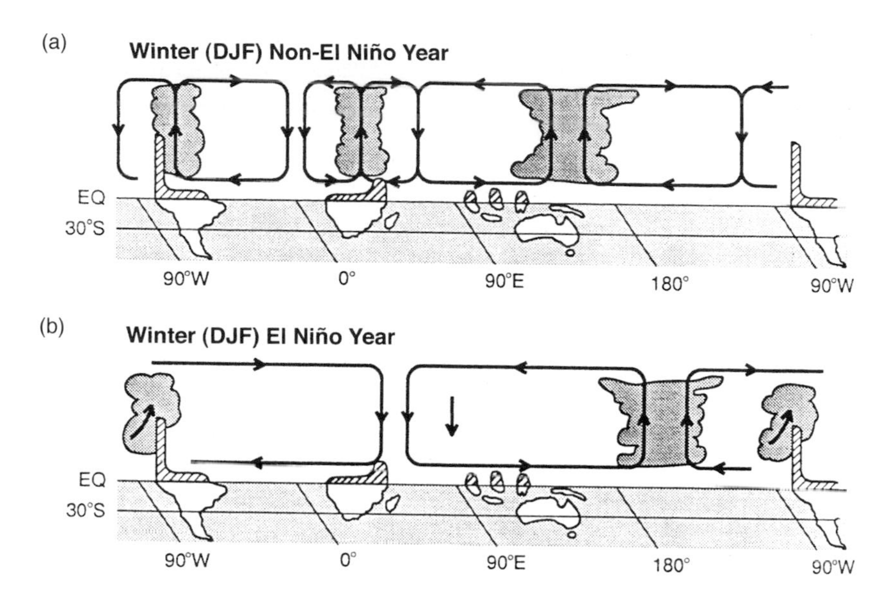


*Figure 4.2. (a) A schematic representation of the Walker Circulation; a cross-section at the equator (EQ) of the atmospheric and oceanic features in the "non-El Niño" phase of the Southern Oscillation. Non-El Niño conditions include normal (around average sea surface temperatures) and La Niña (cold) phases. In some parts of the globe La Niña is viewed as an extreme phase of normal. (b) El Niño (warm) phase of the Southern Oscillation. DJF, December, January and February. (WMO, 1984, p. 15.)*

surface temperatures became unseasonably warmer, and that warming propagated westward toward the central Pacific.

In March 1982, American meteorologists Eugene Rasmusson and Thomas Carpenter wrote a research article that defined the canonical (i.e., typical) El Niño. However, the 1982–83 event that developed later in 1982 differed in several respects from the earlier ones used to construct the canonical El Niño. This prompted El Niño researchers to consider more seriously the differences as well as the similarities among El Niño events.

The canonical image of El Niño, however, may have been misleading, because it was made up by averaging the conditions of several post-World War II El Niño events. Averaging takes away any indication of how weak or strong an El Niño event might be and the ways in which each event differed from an average (stereotype). As scientists later observed, the warming of ocean surface water can occur in the central Pacific first before it warms along the Peruvian coast, the opposite of what appears to have happened in the 1972–73 event.

Now that scientists have the technological tools to look more closely at each specific event, they can identify those features that are similar, as well as those that are unique, to any given event. For example, it is not difficult to see that El Niño events can form in at least two different ways, on the basis of whether the sea surface temperatures first heat up either in the eastern or the central Pacific along the equator. Such tools can help researchers to identify the range of characteristics and impacts of an El Niño event.

El Niño events have a life cycle. Rasmusson and Carpenter (1982) identified several stages of its development: antecedent, onset, peak, transition, maturity, and decay. The components of this life cycle are depicted in Figure 4.3, which was based on the set of El Niño events that took place before 1982. It is the same set of events that they used to define the canonical El Niño. The interesting point about this figure is that it also shows the time and location of the increase in rainfall across the tropical Pacific associated with a developing El Niño event.

With the realization after the 1982–83 event that such episodes can begin at different times of the year, tying the phases of the life cycle to specific months of the year became more of a problem for researchers and especially for forecasters. Sometimes the warming along the Peruvian coast occurs before the warming in the central Pacific, and sometimes it follows it. Even though there are different types of El Niño event based on the timing of onset and location of anomalous increases in sea surface temperatures, they still evolve in a similar way, undergoing the same life cycle sequence of growth to decay.

Australian meteorologist Neville Nicholls characterized the El Niño process, using terms slightly different from those of Rasmusson and Carpenter: a precursor phase, an onset phase, a phase when anomalous conditions grow and mature, and a phase during which those anomalous conditions decay (Nicholls, 1987). These can be described in the following way:

1. Precursor phase: One could argue that the precursor phase for El Niño begins at the end of a cold phase (La Niña) event, when sea surface temperatures have returned to near average (normal).

   Let us assume that the precursor phase begins just after the height of the cold phase. For reasons that are not completely understood, the strong westward-flowing winds (the trade winds) begin to weaken, the sea level in the western Pacific reaches its peak, and the sea level along the western coast of South America reaches its minimum for that cycle. With a weakening of the surface westward-flowing winds, equatorial coastal upwelling begins to reduce, and sea surface temperatures in the central and eastern equatorial Pacific begin to warm up. This is the transition

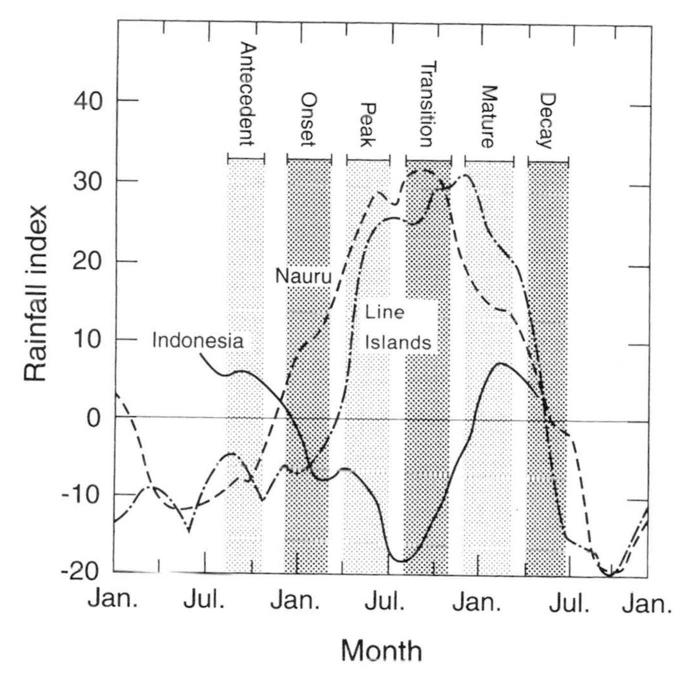


*Figure 4.3. Development of rainfall over time at three locations in the equatorial Pacific Ocean: Indonesia, Nauru (167° E), and the Line Islands (near 160° W) for the El Niño events before 1982. Note the out-of-phase relationship between Indonesian and Nauru rainfall anomalies, and the eastward movement of the rain anomaly from Nauru to the Line Islands. The rainfall tracks the anomalously warm sea surface as it moves eastward into the central Pacific. (From Tomczak and Godfrey, 1994, reprinted by permission of Butterworth Heinemann Publishers, a division of Reed Educational and Professional Publishing Ltd.)*

phase, moving out of a cold event through an average (normal) range and toward a warm event of unknown magnitude.

2. Onset phase: Around December of each year (the beginning of the Peruvian summer), there is a seasonal slackening of the winds off the coasts of Peru and Ecuador. At that time, the upwelling of cold water along the coast slows down, and the surface water heats up, lasting until March or so. If that seasonal warming were to continue into April and May, it is likely that the onset of a warm event of some magnitude would be under way. However, the onsets of some El Niño events (for example, 1982–83) have occurred later in the year (after August), when the sea

surface temperatures had already returned to normal in April after their seasonal warming.

3. Growth and maturity phase: As the months proceed, the sea surface temperatures in the central and eastern equatorial Pacific become increasingly warmer, and upwelling ceases to bring nutrient-rich cold water into the sunlit zone of the ocean surface. The sea level pressure in the South Pacific (near Tahiti) drops, and the pressure at Darwin increases. With the weakening of the westward-flowing winds and the strengthening of the eastward-flowing winds, the area covered by warmed surface water expands in the central and eastern Pacific. Sea surface temperatures can increase by 1 to 4 degrees Celsius or more (as happened in 1982–83). The sea level in the western Pacific drops a few tens of centimeters, while sea level in the eastern equatorial Pacific increases.
4. Decay phase: This phase begins once maximum sea surface temperatures have been reached in the central and eastern equatorial Pacific, and surface temperatures begin to respond to changes in wind speed and direction across the basin. The thermocline depths begin to move in the opposite direction (once again becoming deeper in the west and shallower in the east), and the warm water pool begins to thicken in the western part of the basin. Westward-flowing winds again begin to strengthen and eastward-flowing winds weaken. Coastal and equatorial upwelling begins to strengthen, bringing more cold, deep water to the ocean's surface. And the cycle (i.e., oscillation) begins once again toward the onset of a cold phase. Satellite images confirm the reappearance of biological productivity in the equatorial Pacific especially near the Galápagos Islands.

Scientists often refer to the year of the actual onset of El Niño as year (0), the year before it as year (− 1) and the year following the onset as year (+ 1). The same categorization is used for La Niña years as well. Scientists know how best to interpret this way of looking at various El Niño or La Niña years. However, to non-scientists using it, it can be very misleading. For example, since the mid-1970s there have been twice as many El Niño events as La Niña events. However, sometimes El Niño year (+ 1) can prove to be a La Niña year (0), as was the case in 1997–98. El Niño ended in May 1998 and La Niña began to develop around July 1998. How then should one refer to year 1998? El Niño events can be followed in year (+ 1) by another El Niño. Would that then be labeled year (+ 2) or a new El Niño year (0)? There is also a need to decide how to treat near average (e.g., normal) sea surface temperature years. Thus El Niño year (0) can be followed by El Niño year (+ 1), La Niña year (0) or normal (0). Clearly, more research is needed on how best to identify El Niño and La Niña years. This would make it possible for non-scientists to use El Niño and La Niña information to identify reliable correlations with distant impacts.

A few key changes in Pacific environmental conditions merit close

> Jacob Bjerknes was, of course, the scientist who recognized that the interplay of ocean and atmosphere was responsible for the generation of El Niño. My idea that El Niño is a heat relaxation of the equatorial Pacific ocean expands on his work and emphasizes the importance of ocean circulation in this process. The west Pacific warm pool slowly grows in size and depth, because ocean circulation is not capable of removing all the accumulated heat from this area. As the warm pool increases, atmospheric convection starts to move east toward the central Pacific and the associated winds trigger the equatorial Kelvin wave in the ocean. This wave moves heat eastward [along the equator] and poleward [along the eastern boundary] and thus removes heat from the area of the warm pool. This heat relaxation process determines the length of an El Niño cycle.
>
> Klaus Wyrtki, University of Hawaii, 1996

monitoring: sea surface temperatures, winds, sea level, sea level pressure, changes in the depth of the thermocline, and outgoing longwave radiation. However, processes in the Pacific Ocean and atmosphere and the results of their interactions are quite complicated because each has many components and their impacts in distant locations may also be influenced by changes in other oceans, especially in the Indian Ocean and the region in between these oceans known as the Indonesian throughflow (where water passes between the Indian and Pacific Oceans). While some scientists may favor watching for changes in sea surface temperatures in a specific location as the leading indicator of an El Niño event, others might consider the monitoring of the surface winds to be more important; still others might focus on the Southern Oscillation or the changes in outgoing longwave radiation that suggest where rain is falling.

The El Niño events that scientists had considered to be typical (i.e., the ones that occurred in the post-World War II years until 1982) have differed in several respects from those after 1982. There is not just one type of El Niño. Thus, despite progress made in understanding the El Niño phenomenon in general, it has been difficult for scientists to forecast the precise timing of the onset or of the decay of a warm or cold event, or the intensity of either extreme. It is likely that researchers will eventually identify the existence of various types of El Niño. The more able that scientists are to explain the variations from one El Niño event to the next, the more confidence the potential users of El Niño information will have in using El Niño forecasts in their decisionmaking processes.

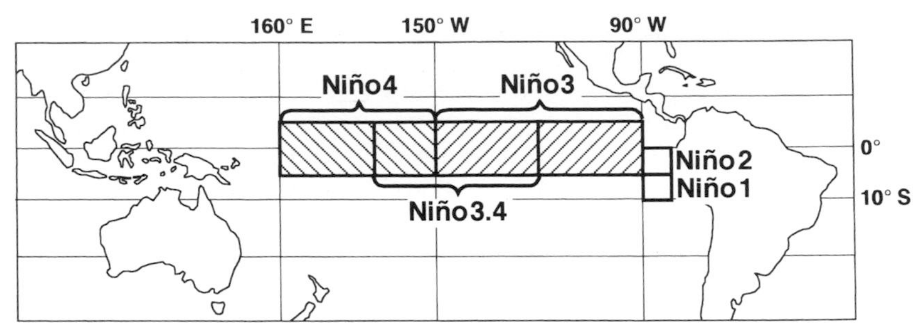


*Figure 4.4. Map showing five regions (referred to as Niño1, Niño2, etc.) in the Pacific identified as important locations for monitoring winds, sea surface temperatures, and rainfall activities, changes that may be associated to varying degrees with the El Niño process.*

## Niño regions

Scientists have identified five regions in the equatorial Pacific that they consider to be worthy of special attention, with regard to the long-term monitoring of El Niño processes. Their locations are shown in Figure 4.4. Researchers are increasingly focusing on environmental changes in the Niño3.4 region in order to identify the onset of an El Niño event.

Each of these Niño regions provide different kinds of information about El Niño, La Niña, normal, or the Southern Oscillation.

- Niño1 is the upwelling region off the coasts of Peru and Ecuador. It is sensitive to changes in the ocean and the atmosphere, both seasonally and especially during El Niño episodes. Coastal upwelling processes in Niño1 are particularly sensitive to changes in air–sea interaction in the central and eastern equatorial Pacific. Data are collected a couple of times per year by cruises of research vessels in these coastal waters. There are no moored buoys to collect data in this region, although there are plans to add some of them to monitor this coastal region.
- Niño2 represents the region around the Galápagos Islands in the equatorial Pacific. Equatorial upwelling processes in this area are also sensitive to seasonal, as well as El Niño-induced, changes in the marine environment. Niño2 is a transition zone between the central and eastern equatorial Pacific, sensitive to changes in either Niño1 or Niño3, or both.
- Niño3 is in the central equatorial Pacific, where there is a large El Niño signal but not much sensitivity to seasonal changes in air–sea interaction. Information on changes in surface winds in this region has been used by Mark Cane and Stephen Zebiak to forecast the onset of El Niño events. According to Cane (1991, pp. 357–8), "a warming in this region is thought

to influence the global atmosphere strongly. It is probably the best single indicator of an ENSO episode likely to affect global climate".

- Niño3.4 is a relatively new region in the tropical Pacific increasingly used by more researchers to correlate changes in sea surface temperatures and surface winds there to climatic anomalies around the globe. Many researchers now use changes in Niño3.4 instead of Niño3 in their El Niño forecast modeling activities. It overlaps the Niño3 and Niño4 regions, as shown in Figure 4.4.
- Niño4 encompasses part of the western equatorial Pacific known as the warm pool. Here, sea surface temperatures are the highest in the Pacific. During an El Niño event, there is a relatively small change in sea surface temperatures (a cooling). However, that small change in temperature is important, because the warmest water at the ocean's surface and the cloud-producing processes that tend to follow it move away from the western Pacific toward the central and eastern Pacific. Hence during El Niño there are dry conditions in several countries in the western Pacific and very wet conditions in northern Peru and southern Ecuador.

## Waves *in* the ocean

Hidden from the naked eye, waves exist inside the ocean, several meters to hundreds of meters below the ocean surface. These are referred to as internal waves. No matter how hard I might try to avoid the mention of these waves, a reliable understanding of El Niño cannot be gained without it. Scientists have identified two types of internal wave: the Kelvin wave and the Rossby wave.

*Kelvin waves* are created by winds blowing over the Pacific Ocean surface from the west toward the east along the equator. Before a warm event develops, eastward-flowing wind bursts increase over the area of warm sea surface temperature (the warm pool) to the east of New Guinea. These so-called westerly wind bursts produce two effects: they move warm water eastward from the warm pool toward the central equatorial Pacific, generating the warm sea surface temperatures observed there, and they also produce Kelvin waves, causing a deepening of the thermocline and an increase in sea surface temperatures in the eastern Pacific. Changes in the thermocline cause small changes in sea level (which can be measured by satellite). Although the thermocline cannot be measured directly by satellite, satellites can measure, as proxy information, changes in sea level height that have been correlated with changes in the thermocline. Kelvin waves can change the depth of the thermocline by 30 meters or more, and the sea level by tens of centimeters. More specifically, this decreases the amount (i.e., volume) of warm water in the western Pacific's warm pool, as the thermocline becomes shallower than normal, while, in the eastern part of the basin, the volume of warm water increases. Kelvin waves are the mechanism that is said to be responsible for the popular notion that warm water "sloshes" back and forth (from west to east and back again) across the equatorial Pacific basin, as water does in a bathtub each time the person in the tub moves (Harrison and Cane, 1984).

The net effect of a series of Kelvin waves is to raise the sea surface temperature systematically over much of the equatorial Pacific. The formation of cloud systems and thunderstorm activity, following the warm water, and usually located over Indonesia, the Philippines, northeast Australia, Papua New Guinea and the warm pool, also moves eastward into the central and eastern Pacific. This weakens the trade winds even more as the Walker Circulation moves eastward, causing even further warming of the sea surface temperature, and so on. Scientists call this interaction a "positive feedback" or reinforcing mechanism between the atmosphere and ocean, one that finally results in El Niño conditions.

A Kelvin wave takes about $2\frac{1}{2}$ months to travel across the Pacific basin, a distance of one-third of the circumference of the Earth. Once forced to begin, Kelvin waves move eastward, independent of the season. Although they have been centrally implicated in starting off an El Niño, every Kelvin wave does not always lead to an El Niño event.

When the Kelvin wave "hits" the coast of South America, it generates coastal Kelvin waves that propagate both to the north and to the south along the western coasts of North and South America, respectively. It also generates what is known as a *Rossby wave*, a westward-moving internal wave that travels at one-third the speed of a Kelvin wave. It takes about 9 months for a Rossby wave to cross the Pacific. Rossby waves depress the thermocline in the western Pacific region. Some people think that the creation of a Rossby wave begins the process of decay of an El Niño, suggesting that the onset of an El Niño carries with it the seeds of its own destruction. While such internal waves are not visible to the naked eye, there are ways to identify their existence using indirect measures, shown using satellite images in Figure 4.5 and as suggested in Figure 4.6.

## A lingering El Niño?

In the mid-1990s, some oceanographers claimed to have discovered that the impacts of El Niño events on the oceanic environment did not dissipate within a few years, as everyone had believed. They suggested that the consequences of the 1982–83 El Niño were still being felt in the northwestern Pacific Ocean as late as in 1994 in the form of a Rossby wave. This very slow-moving wave, according to researcher Gregg Jacobs and his colleagues (1994), affected the location of the northern extension of the Kuroshio Current, a warm current in the Pacific that functions similarly to

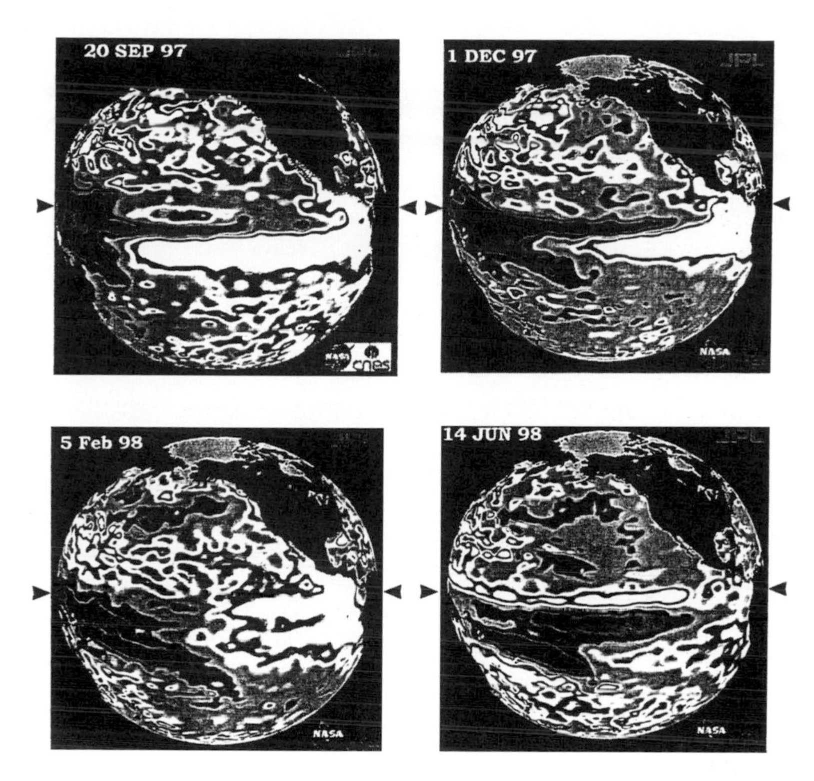


*Figure 4.5. Satellite images of sea level differences. The dark zone along the equator represents a drop in sea level and the light zone reflects an increase in sea level. These changes are used to track the growth and decay of an El Niño over time. (Photos from NASA JPL.)*

the warm Gulf Stream in the North Atlantic. The Kuroshio Current determines, to a large extent, the seasonal climates of the countries bordering the North Pacific. They suggested that this "lingering" Rossby wave pushed the warm Kuroshio Current further to the north, affecting ocean temperatures and perhaps weather patterns on the North American continent. While this view was received with interest by some researchers, it remains speculative, highly controversial, and difficult to imagine. The Rossby wave was there; however, its alleged influence remains unclear.

## Cold events

Warm events (El Niño) are only part of the cycle taking place with regard to changes in Pacific sea surface temperatures. Cold events (La

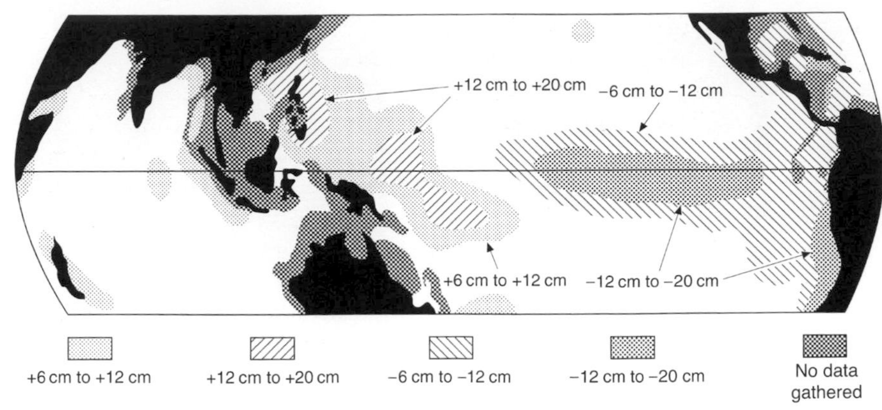


*Figure 4.6. Changes in sea level between 1987 and 1988 (1988 data minus 1987 data). Scientists use maps like this to show sea level changes in the Pacific between an El Niño year (a warm event) and a La Niña year (a cold event). This map shows the recovery of the sea level from the 1986–87 El Niño into the opposite phase (sea level drops in the east from being high in 1987 (El Niño) to being low in 1988); in the west it goes from being low in 1987 to being high in 1988. This is like the seesaw pattern one sees with the Southern Oscillation. This map also shows that the drop in sea level in the east is trapped near the equator and along the coast (where it moves poleward in each hemisphere). The equatorial and coastal signal is characteristic of equatorial Kelvin (internal) waves. It is also interesting to note the connections to higher latitudes implied by positive sea level changes in the western part of the Pacific. (After Koblinsky* et al., *1992.)*

Niña) are also part of the cycle, with warm and cold events usually but not always appearing at the extremes of the seesaw pattern of sea level pressure, the Southern Oscillation. It is, therefore, essential to view El Niño and La Niña together as part of the same phenomenon – the ENSO cycle.

During cold events, sea surface temperatures in the eastern and central Pacific decrease by a few degrees Celsius from the long-term average. Today, researchers argue that cold events produce weather and climate anomalies in distant locations that are in general opposite to those produced in the same locations by El Niño. For example, cold events are believed to be associated with good rains and favorable agricultural production in Indonesia, Australia, and northeast Brazil, whereas El Niño has been associated with droughts in these regions.

Aside from meteorologists Harry Van Loon and, later, George Kiladis, relatively few researchers have shown continued interest in cold events. Another exception has been oceanographer James O'Brien, at Florida

State University, who was among the first to popularize La Niña's impacts, especially in the southeastern USA. Perhaps one reason for the lack of interest in cold events is because such events are associated with periods of weather and climate conditions that are *perceived* as normal in various regions. According to one definition of La Niña, only one cold event occurred between 1975 and 1988. Other La Niña events have since occurred in 1995–96 and 1998–2000. Their relative infrequency since the mid-1970s has most likely also served to reinforce the lack of interest in cold events and their potential consequences.

From the perspective of societal impacts, it also appears that for some countries a cold event, or, more precisely, any situation that is not deemed to be an El Niño, has generally been viewed as "normal" tropical Pacific conditions. This view, at least to the public, has been strongly reinforced by such media statements as "El Niño is ending and the weather will return to normal." And, until recently, researchers generally referred either to "El Niño" or to normal non-El Niño conditions.

# 5 The biography of La Niña

## Background

For millennia, people have tried to understand, predict, forecast and even guess at what the natural variations of local and regional climate might be on seasonal and year-to-year time scales. Aristotle's *Meteorologica*, written in the fourth century BC, was perhaps among the earliest recorded treatises on weather and climate. Weather and climate have been of concern to individuals, groups, and governments for a variety of reasons – for food production, water resources management, public safety, passing interest, curiosity, personal comfort, public welfare, and even survival. At the end of the nineteenth and the beginning of the twentieth centuries, weather and climate forecasting had started to become a growing research "industry".

As noted earlier, efforts in the late 1800s and early 1900s to predict the behavior of the monsoon on the Indian subcontinent sparked several attempts to forecast the breakdown of the monsoon in order to avert recurrent food crises and famines. Scientists have since developed a Southern Oscillation Index (SOI) to forecast droughts in Australia and elsewhere. The SOI also served as an indicator of the onset of El Niño (and of La Niña as well) along with changes in Pacific sea surface temperature (SST). With each successive El Niño event after 1972, the scientific community has become increasingly interested in the phenomenon, that interest and research reaching new heights in the late 1990s.

While the 1972–73 event and its impacts were noteworthy, the 1982–83 El Niño and its impacts commanded the serious attention of the scientific community and governments around the world. However, neither of these El Niño events was predicted by forecasters and blind-sided scientists and societies worldwide. The intense 1997–98 El Niño demonstrated to a wide range of potential users of climate-related information the value of El Niño-related information for their specific needs. The increase over the past few decades of interest in El Niño is noteworthy, but it had overshadowed the importance of other parts of the ENSO cycle that had

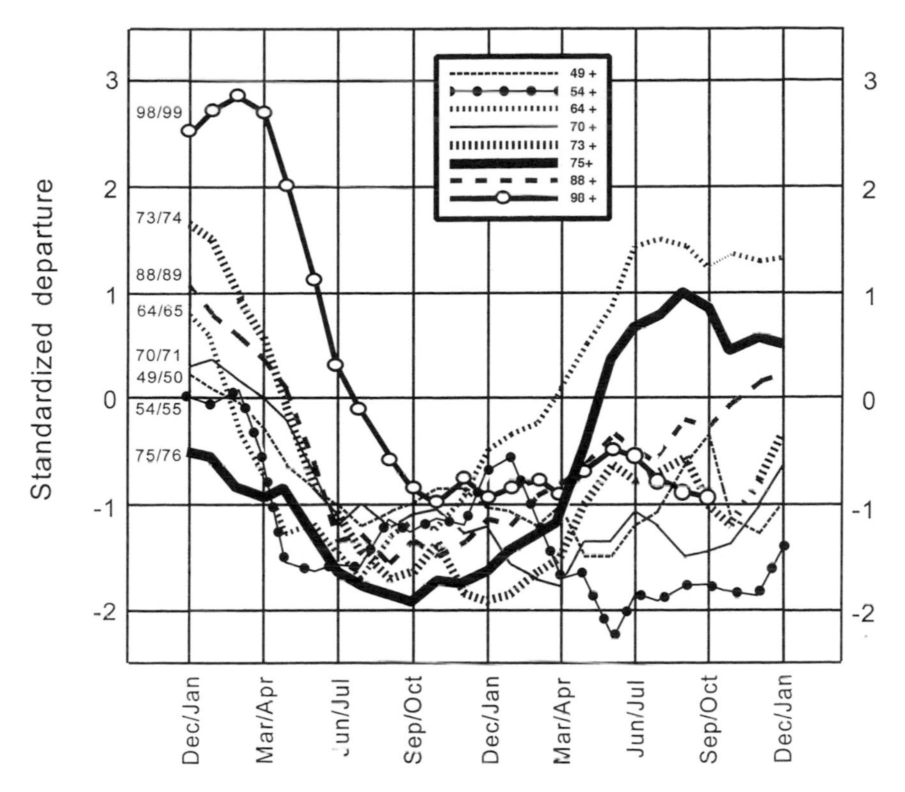


*Figure 5.1. Multivariate ENSO Index (MEI) for the seven strongest historic La Niña events since 1979 versus conditions as of November 1999. MEI is based on the six main observed variables over the tropical Pacific. These six variables are: sea level pressure, zonal and meridional components of the surface wind, sea surface temperature, surface air temperature, and total cloudiness fraction of the sky. (Adapted from Climate Diagnostics Center's website (www.cdc.noaa.gov/~kew/MEI/mei.html).)*

been neglected by scientists, policymakers and the media. However, the rapid decay of the 1997–98 El Niño in May 1998 changed all that. It sparked researchers to forecast the eventual onset of a major La Niña event. A La Niña did develop, as shown in Figure 5.1.

The media have referred to La Niña in many ways, calling it at one time or another one of the following: a cold episode, a cold event, a cold phase, a mature cold episode, El Viejo, the cold phase of ENSO, El Niño's counterpart, the cold counterpart of El Niño, El Niño's sister, the girl child, anti-El Niño, non-El Niño, seasons with cold SSTs, an anomalous cooling, the opposite of El Niño, a sisterly event, El Niño's cold water opposite, the

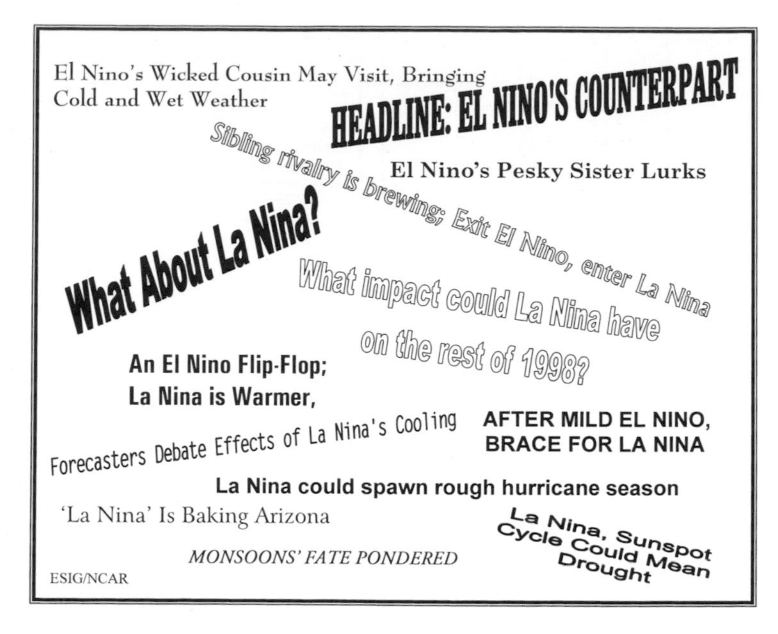


*Figure 5.2. La Niña headlines taken from around the world. (From ESIG/NCAR.)*

flip side of El Niño, abnormally cold, El Niño's lesser known twin, a periodic abnormally cold sea surface current, the other extreme of the ENSO cycle, and so on. A selection of media headlines are shown in Figure 5.2.

Speculation aside and with a few notable exceptions, there has been a relative dearth of *focused* research on the cold extreme of the ENSO cycle (i.e., La Niña) when compared with El Niño coverage. Part of the reason for this relative lack of interest may be the fact that only a few cold events have occurred in the 25 years that preceded the 1998–2000 event. However, the number of La Niña events is determined by each researcher's definition of La Niña. As yet, there is no general agreement among scientists on a single definition and, therefore, no agreed-upon list of cold event years. Another reason for the lack of attention paid to La Niña may have been the perception by government funding agents that the societal impacts of La Niña are not as devastating as those of El Niño. This perception was recently challenged by meteorologist Stanley Changnon (1999). Changnon's article sparked a debate about which extreme – warm or cold – has the worst impacts for North America (e.g., Associated Press, 1999).

Yet another reason for relatively less attention having been paid to La Niña by scientists (again, except for Harry van Loon, James O'Brien, and George Kiladis) is the way that Pacific SSTs have been classified as falling

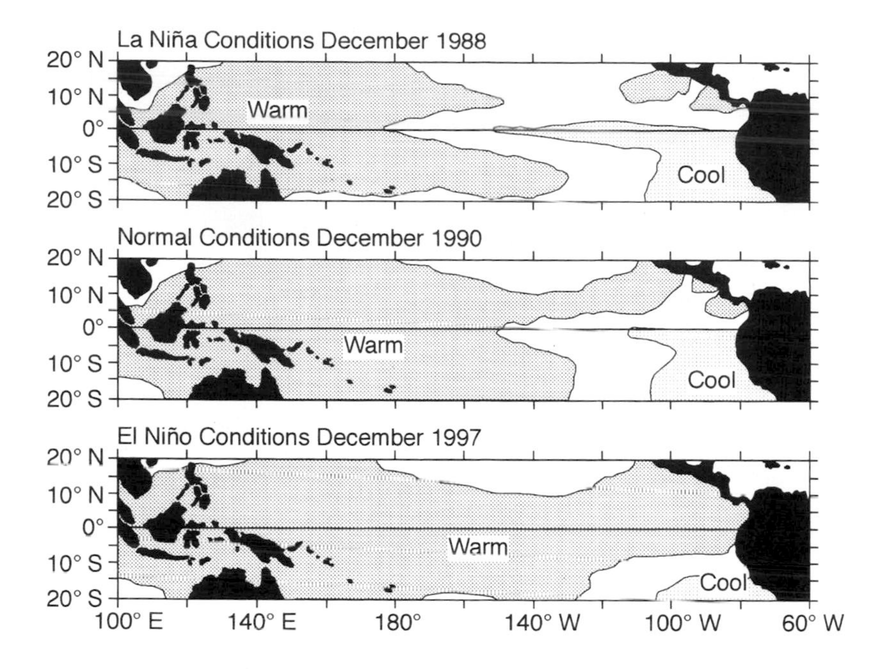


*Figure 5.3. The three phases of sea surface temperatures in the tropical Pacific Ocean. Redrawn from NOAA's Pacific Marine Environmental Laboratories (PMEL) website. www.pmel.noaa.gov of December 1998.)*

into either one of two possible states: El Niño and non-El Niño, with the dividing line between them being a statistically determined average SST condition in the central Pacific Ocean. To policymakers in some of the countries that are severely affected by El Niño, changes in Pacific SSTs that are *not* El Niño conditions are perceived to be normal, even if the expected good rains for agriculture become torrential rains and cause flooding.

However, La Niña, like El Niño, represents only one extreme end of the spectrum (or oscillation) in SSTs. One could then argue that there really are three possible states for equatorial Pacific SSTs – a range of extreme cold, a range of extreme warm, *and* a range of SSTs around average, referred to as normal. Those who look at the Pacific SSTs as having only two states tend to consider "normal" only as the point of change from one extreme to the other. Figure 5.3 shows SST conditions for the three phases and suggests why, until recently at least, little distinction had been made between La Niña and normal conditions.

As the 1997–98 El Niño began to decay rapidly in May 1998, the sea's surface and subsurface temperatures dropped, passed through average,

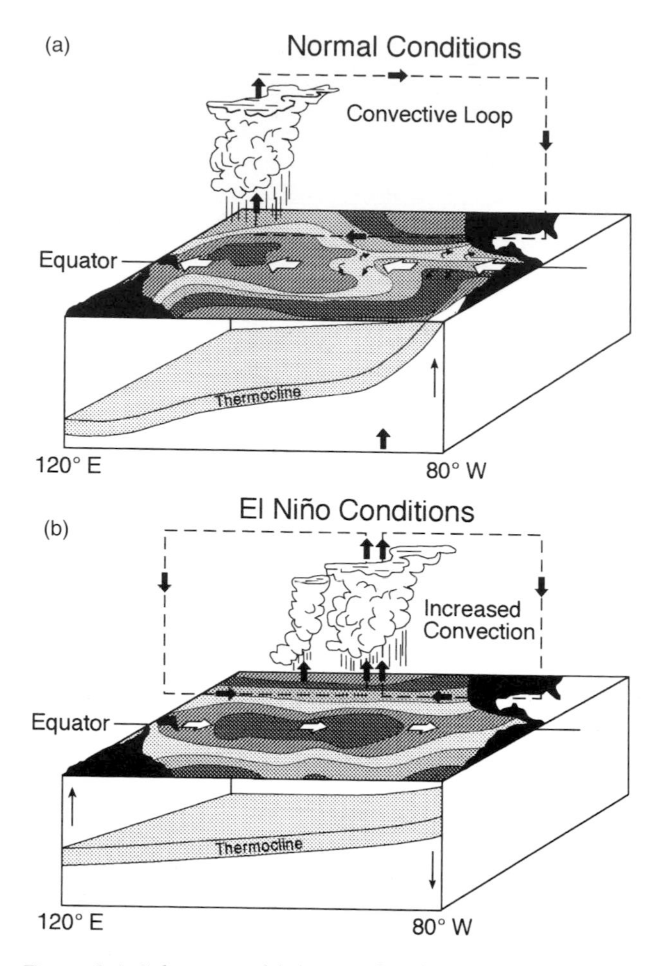


*Figure 5.4. Schematic of (a) normal and (b) El Niño conditions in the equatorial Pacific. (From McPhaden* et al., *1998.)*

and moved toward a cooling at the rate of about 1 degree Celsius per week over a 4–6 week period. This indicated the likely onset of a La Niña and captured the attention of scientists and media throughout the world. A belief prevailed at the time that such a sharp drop in SSTs (not witnessed in previous decades), following the peak phase of a strong El Niño, meant that a strong La Niña was about to develop. At least for a few months, La Niña captured news headlines around the globe.

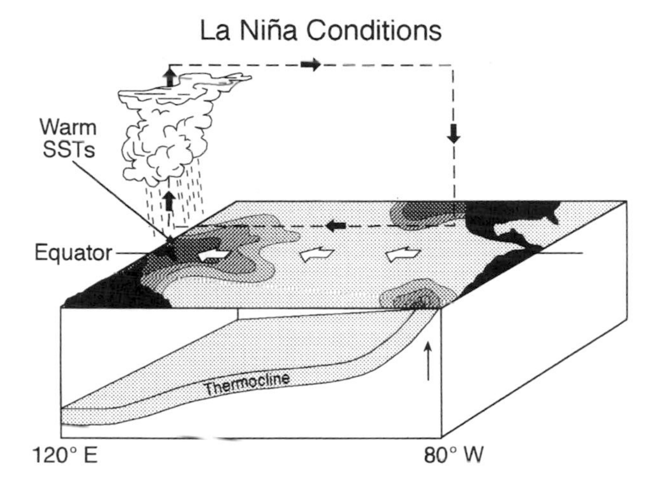


*Figure 5.5. Schematic depicting La Niña conditions in the equatorial Pacific. SST, sea surface temperature. (From ESIG/NCAR.)*

*What is La Niña?*

As noted earlier, the warm pool of water that is usually present in the western equatorial Pacific provides moisture to the atmosphere through evaporative processes that lead to rain-producing cloud formations. As a result, heavy rains (considered to be normal in that region) provide the water resources needed in East and Southeast Asia and in Pacific island nations for use in agriculture, hydropower, navigation, and drinking water. In the eastern part of the Pacific, when the upwelling of extremely cold deep water occurs along the equator, arid conditions are enhanced along the western coast of South America. This trans-Pacific basin situation is considered to be a *non-El Niño* condition, or normal. Non-El Niño conditions have frequently been viewed as including both normal and La Niña conditions. The schematics depicting the ENSO process, drawn by scientists before the 1997–98 El Niño and the 1998–2000 La Niña, showed only two states of air–sea interaction in the tropical Pacific – normal and El Niño conditions (Figure 5.4).

Most recently, however, they have produced a La Niña schematic that is somewhat different from the one representing normal, as shown in Figure 5.5.

*Physical indicators of La Niña*

La Niña can be said to exist when *extreme* cold sea surface temperatures appear in the central and eastern equatorial Pacific for an extended period of time (several months), when westward-blowing winds along the equator are strong, when the value of the SOI is highly positive, when the thermocline in the western Pacific is depressed and in the eastern Pacific is near the ocean's surface, and when the sea level in the western Pacific drops and in the eastern Pacific rises. As with El Niño, La Niña can develop to varying levels of intensity: weak, moderate, strong, very strong, and extraordinary. However, we have not witnessed enough La Niña events in the past few decades to use these categories with any degree of reliability. The intensity or duration of La Niña (as for El Niño) is difficult to forecast at present and can be determined primarily through monitoring its development over time.

Despite the relative lack of interest in La Niña, statements alluding to the global impacts of La Niña can be found in the scientific literature of the past two decades. Most of the impacts maps that exist in print and on the Internet are variations of those developed about 15 years ago by Chester Ropelewski and Michael Halpert (1987). However, since then several El Niño and La Niña events have occurred. Clearly, it is time to review and, if necessary, update such impacts maps. Given the widespread use of these maps worldwide by forecasters, ENSO researchers, policymakers, the media and the public, updated ENSO maps would enable these forecast users to prepare for the impacts of ENSO extremes that they might reasonably expect in their regions.

Aside from these generic impacts maps, there are other La Niña-related issues that are in need of further investigation, including, but not limited to, the following: the degree of symmetry between La Niña's physical characteristics and those of El Niño, improved methods for the attribution of societal impacts to La Niña events, the physical and statistical teleconnections that can legitimately be linked to (or blamed on) La Niña events, identifying the needs for monitoring La Niña as opposed to El Niño, the actual state of the art of forecasting ENSO's extreme events, and the possible influence that global warming might have on the ENSO process (see Glantz, 1998b).

### *What is La Niña's relationship to El Niño?*

Most articles suggest that El Niño and La Niña are opposites, using phrases such as "La Niña is the other extreme of the ENSO cycle"; "As El Niño goes away, La Niña starts to rise"; "Like a mirror image of El Niño, it produces extreme weather and abnormal conditions in the western Pacific regions opposite to those El Niño produces there," and so on.

From such headlines, it seems that the occurrence of a La Niña has been considered to have been dependent on El Niño. The opposite, however, is not perceived to be the case. Whereas most El Niño articles do not refer to

La Niña, most articles about La Niña make some explicit reference to El Niño. At least until the end of the 1990s, La Niña events have clearly been living in the shadow of El Niño. Recall that El Niño was first identified as a problem and named in the early 1890s by Peruvians, whereas the term "La Niña" was first coined about 90 years later in 1985 by Princeton University oceanographer George Philander (1990). Newly heightened awareness of both of the ENSO extremes will probably rectify in future years this imbalance in attention between warm and cold extremes.

Clearly, with the rapid decay of the 1997–98 El Niño and with a growing, but new, interest in the other parts of the ENSO cycle, La Niña has developed its own identity. North America's media, including USA Today, CNN, ABC, NBC, MSNBC and CBS, began to showcase La Niña stories in mid-1998, once forecasters had become convinced that El Niño had begun to decay; "With El Niño on the decline, can La Niña be far behind?" The media reported comments by researchers that La Niña "tends to bring worse weather than does El Niño," citing, for example, the statistics that the number of Atlantic hurricanes increases during La Niña.

Media in other countries have also begun to take La Niña seriously. For example, news media in various countries in Southeast Asia produced banner headlines on La Niña in the middle of 1998. They served to prepare the public for the 1998–2000 La Niña and its regional adverse societal impacts. As another example, media in the Philippines, Malaysia and Brunei actively reported on La Niña, which was expected to bring them heavy rains. In fact, decisionmakers in the countries of Southeast Asia came to realize that their countries can be adversely affected by both extremes of the ENSO cycle. At least in that part of the world, La Niña is becoming a household concern.

### *Which extremes can be blamed on La Niña?*

Societal contributions to atmospheric greenhouse gases through the burning of fossil fuels (coal, oil, and natural gas), tropical deforestation and the use of fertilizer ($NO_x$) and refrigerants (chlorofluorocarbons or CFCs), have been linked to a global warming of the atmosphere. In the summer of 1988, a major drought took place in the US Midwest, which has since been referred to as the most expensive natural disaster in US history. Some researchers quickly blamed the severity of that drought on human-induced global warming. James Hansen (1988), for example, suggested that the effects of global warming on regional and local climates would become more frequent as well as more visible in the near future. The 1988 drought, he argued, was consistent with what one might expect from global warming. At that time, however, an equally plausible hypothesis was proposed by atmospheric scientist Kevin Trenberth: the Midwest drought was a result of La Niña conditions in the equatorial Pacific (Linden, 1988).

Was the 1988 Midwest drought really produced by prevailing La Niña conditions thousands of kilometers away? If that were the case, then there would have been a good chance that a major Midwest drought would accompany the 1998–2000 La Niña. However, can such a conclusion be made with confidence? While there is some evidence that a La Niña summer in North America is likely to be hotter and drier than normal, there is not enough hard evidence to make that fairly specific geographic teleconnection with certainty.

Care must be used in identifying previous La Niña (or El Niño) events that are to be used as analog years. Because there have been relatively few La Niña events in the past 25 years, we do not know the full range of ways that La Niña might affect regional climates in different parts of the world. Identifying a specific La Niña year from the historical record that might be considered to be similar to an impending La Niña year raises expectations about the increased likelihood of a repeat of the societal impacts that occurred during those previous years. If the selection of an analog year is wrong, however, then those expectations about potential damages would have been false expectations, because these damages are not likely to occur.

Identifying La Niña years (and months) is very important for those who look for La Niña analogs to forecast with some degree of reliability the impacts that might occur and to develop strategies to cope with the societal impacts of the ENSO cycle. Although it is very important for users of El Niño forecasts to know with certainty which years were El Niño years, scientists appear to be less concerned about the need to do so. They favor using the long time series of sea surface temperatures, sea level pressure, thermocline depth and outgoing long-wave radiation to identify the ENSO warm or cold events. Nevertheless, the media, policymakers, the public and even many researchers continue to label whole calendar years as either El Niño or La Niña years. Yet, El Niño events can end early in the same year that a La Niña event begins. The reverse is also true. Table 5.1 is a chart produced by Trenberth (1997), giving the months from onset to end of La Niña events, according to his definition of La Niña. The chart clearly shows that labeling whole years as El Niño or La Niña in the absence of other information is misleading.

Therefore, the scientific community needs to do a better job of labeling years as La Niña, El Niño or normal so that researchers and managers in other climate-related fields (agriculture, energy, water, public safety, public policy, etc.) can better prepare at the local level for ENSO-related anomalous extreme climate-related events.

Table 5.1. *La Niña events*

| Begin | End | Duration (months) |
|---|---|---|
| Mar. 1950 | Feb. 1951 | 12 |
| June 1954 | Mar. 1956 | 22 |
| May 1956 | Nov. 1956 | 7 |
| May 1964 | Jan. 1965 | 9 |
| July 1970 | Jan. 1972 | 19 |
| June 1973 | June 1974 | 13 |
| Sept. 1974 | Apr. 1976 | 20 |
| Sept. 1984 | June 1985 | 10 |
| May 1988 | June 1989 | 14 |
| Sept. 1995 | Mar. 1996 | 7 |

*Source:* Trenberth, 1997.

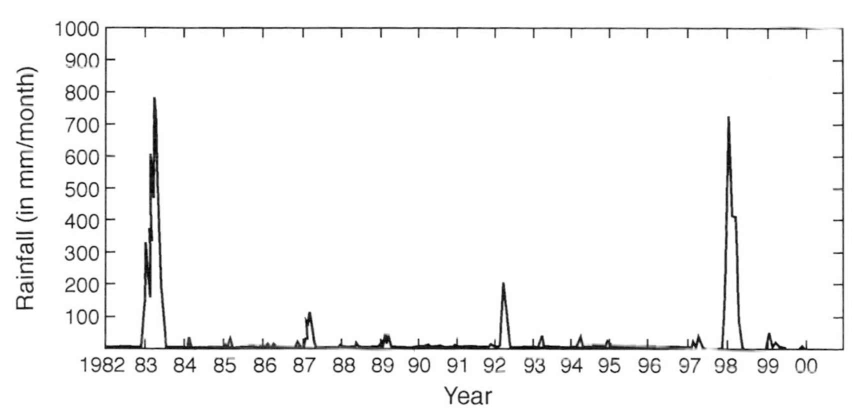


*Figure 5.6. Rainfall for Piura, Peru. (R. Woodman, Instituto Geofisica del Peru (March 1998), updated by N. Ordinola, University of Piura (November 1999).*

## *Suggested probable impacts of La Niña*

A rule of thumb that researchers have created is that the impacts of La Niña are generally the opposite to those of El Niño. For example, droughts tend to accompany El Niño events in Australia, Indonesia, and the Philippines, whereas heavy rains and flooding tend to accompany La Niña in these locations. Southern Africa tends to be drought-plagued during El Niño, but very wet during La Niña episodes. As another example, the coastal zone of northern Peru is arid during La Niña episodes but becomes flood prone during El Niño events, as shown in Figure 5.6.

The actual worldwide impacts of cold events depend on the intensity of the particular La Niña. Some researchers have suggested that the 1988–89 La Niña was a strong one and the 1995 event was weak. Others have suggested that the 1988–89 La Niña was only moderate and that the La Niña in 1984–85 was weak. Much more research needs to be done on how best to classify the intensity of La Niña events. The composite maps shown in Figure 5.7a,b, based on those produced by Ropelewski and Halpert (1987), provide a "*statistically based* generalization" of the *potential* impacts of ENSO cold extremes. By noting the months in which impacts are likely to occur, this map is much more useful than the maps that identify only the generalized locations of La Niña's impacts.

Some maps, such as those in Figure 5.7, are composites of the meteorological impacts of several events, while other maps represent the impacts of an individual La Niña (Figures 5.8a,b, 5.9a,b and 5.10). Still other maps suggest a wide range of anomalies and extreme events that occurred during the 1988–89 and the 1999–2000 La Niña events. While these maps depict many of the extreme impacts that occurred in those years, several of those impacts were not really attributable to La Niña conditions in the Pacific. Although each of these types of impacts map has its benefits, each also has its drawbacks. They must be used with care.

A composite map represents the average of a range of impacts in a given region of several (often an unspecified number of) cold or warm events. The intensity of the impacts of very strong events becomes muted by including them alongside several events of lesser intensity.

A single-year La Niña impacts map includes much detailed information about weather anomalies in a given year. However, many of those anomalies may not have been connected to La Niña. Even with normal SSTs in the central Pacific, there are sure to be extreme meteorological events around the globe. Thus a La Niña (or El Niño) single-episode impacts map must be viewed as illustrative and suggestive, but should not be used as a benchmark map of what *will* happen during any specific cold (or warm) event.

### *Forecasting La Niña*

The primary methods used to forecast the impacts of a La Niña event (of uncertain intensity) are (a) computer model outputs or (b) quantitative or qualitative analyses of past La Niña events and their environmental and societal impacts. These are the approaches of choice by scientists to identify the possible consequences of an ENSO extreme cold or warm event. However, less scientific ways to get a glimpse of the impacts that a La Niña might bring to a given region include projections based on what is likely *not* to occur under La Niña or "normal" tropical Pacific SST

(a)

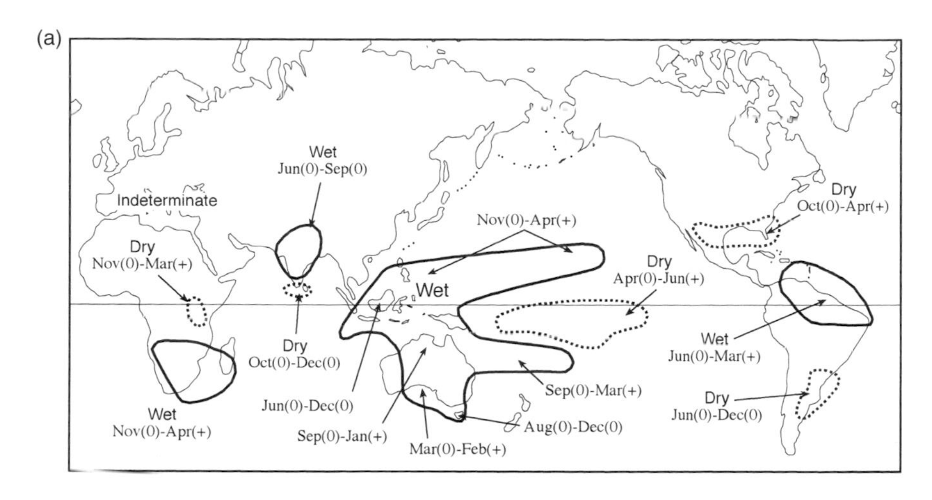


(b)

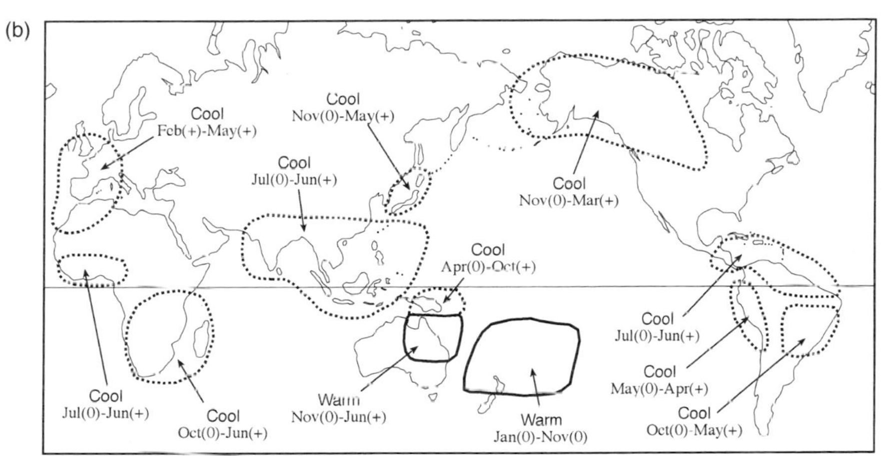


*Figure 5.7. (a) Potential rainfall impacts from La Niña events, and (b) potential temperature impacts from La Niña events. (0) Year of La Niña onset; (+) year following La Niña onset. (After Ropelewski and Halpert, 1987.)*

conditions. One could identify *El Niño* teleconnections considered to be very reliable and then assume that, in the absence of an El Niño, there would be a much lower chance for those El Niño-related anomalies to occur.

For example, the El Niño-associated extreme drought situations in Indonesia, Papua New Guinea and in Australia or the forest fires in Borneo (Indonesia) are much less likely to occur during a La Niña event. So, while

(a)

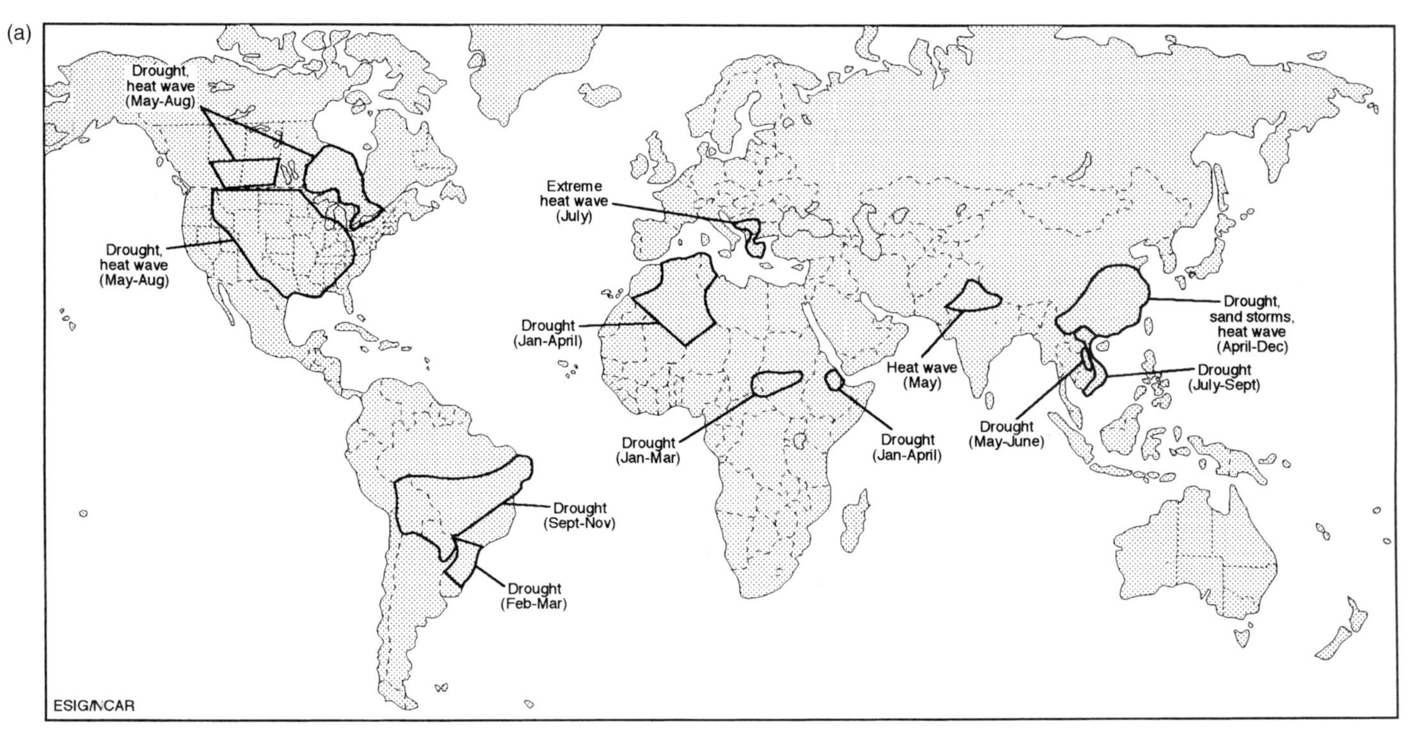
Drought, heat wave (May-Aug)
Drought, heat wave (May-Aug)
Extreme heat wave (July)
Drought (Jan-April)
Drought (Jan-Mar)
Drought (Jan-April)
Heat wave (May)
Drought, sand storms, heat wave (April-Dec)
Drought (July-Sept)
Drought (May-June)
Drought (Sept-Nov)
Drought (Feb-Mar)
ESIG/NCAR

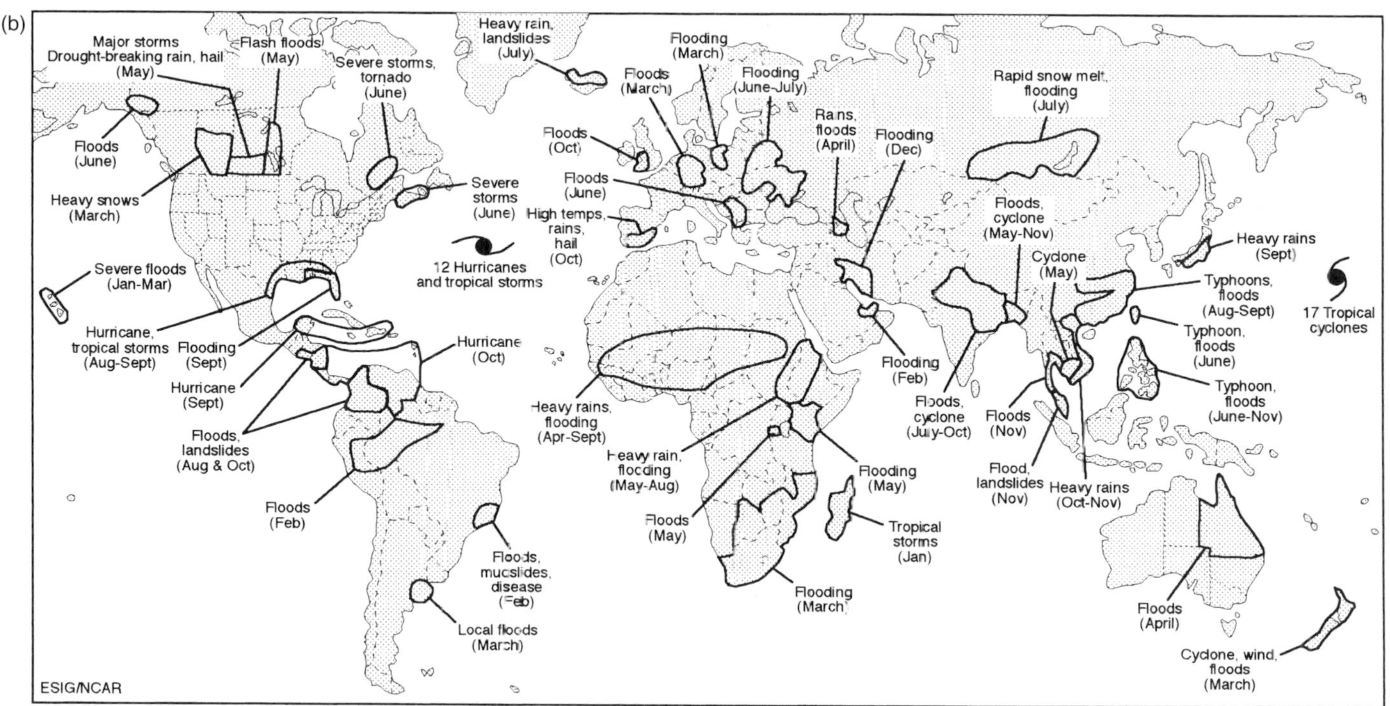


*Figure 5.8. According to Trenberth (1997), a La Niña began in May 1988 and ended June 1989. An El Niño began in August 1986 and ended in February 1988. (a) 1988 droughts; and (b) 1988 floods, rains, and severe storms.*

(a)

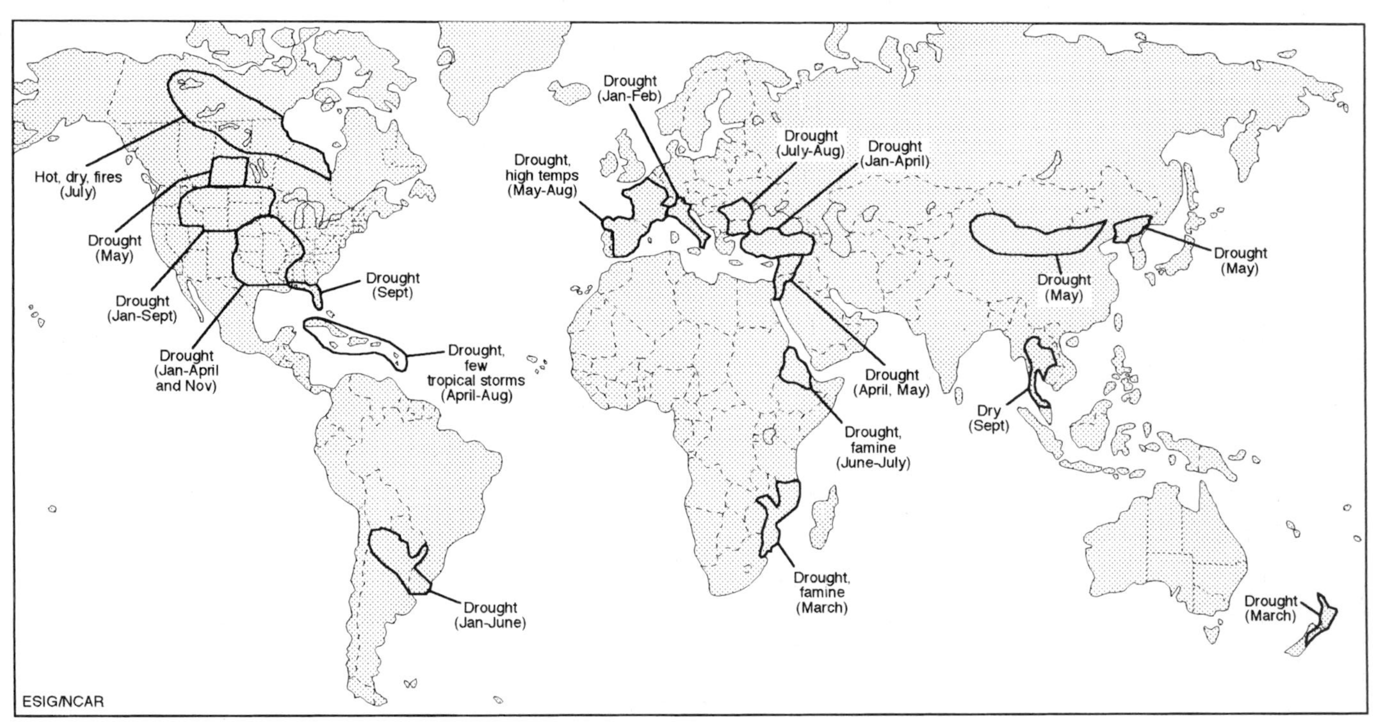

Hot, dry, fires (July)
Drought (May)
Drought (Jan-Sept)
Drought (Jan-April and Nov)
Drought (Sept)
Drought, few tropical storms (April-Aug)
Drought (Jan-June)
Drought, high temps (May-Aug)
Drought (Jan-Feb)
Drought (July-Aug)
Drought (Jan-April)
Drought (April, May)
Drought, famine (June-July)
Drought, famine (March)
Drought (May)
Drought (May)
Dry (Sept)
Drought (March)
ESIG/NCAR

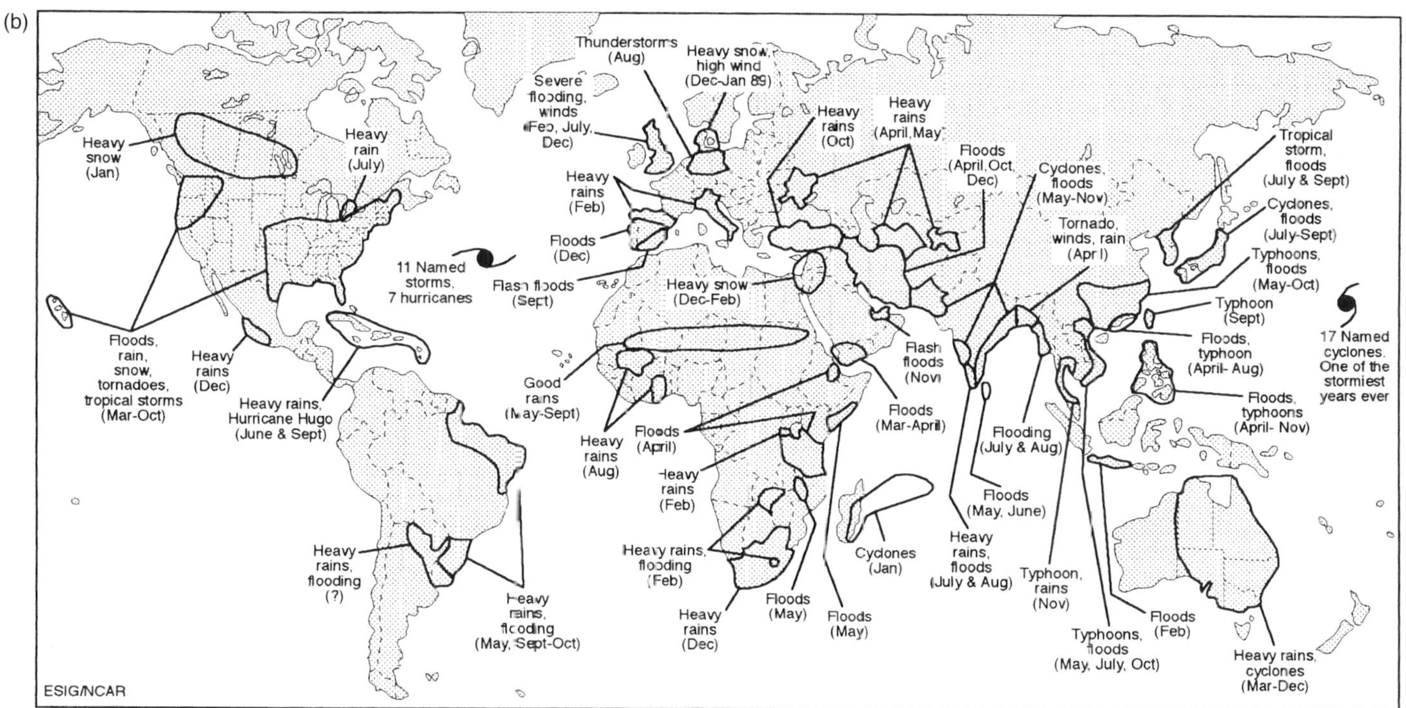


*Figure 5.9. According to Trenberth (1997), a La Niña began in May 1988 and ended June 1989. An El Niño began in August 1986 and ended in February 1988. (a) 1989 droughts; and (b) 1989 floods, rains, and severe storms.*

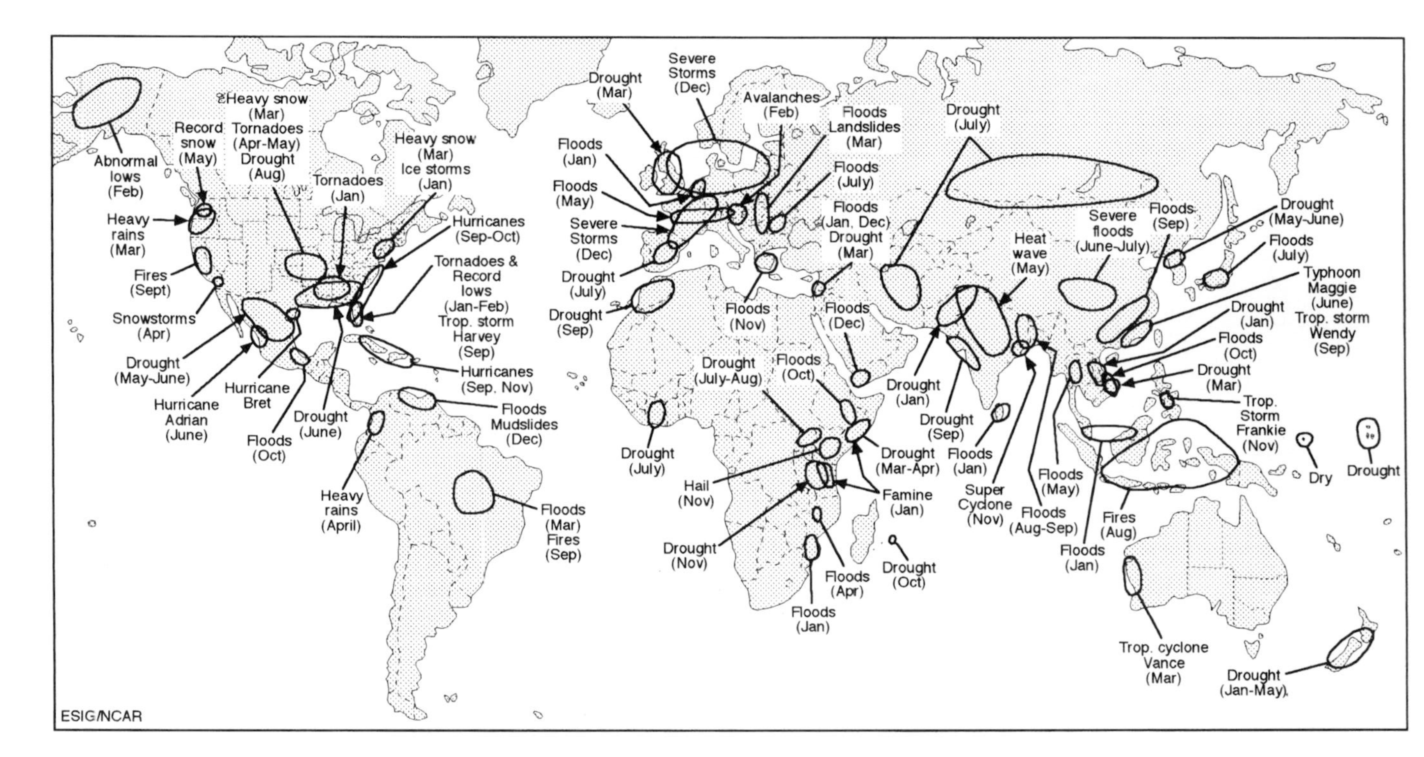


*Figure 5.10. 1999 weather anomalies. Not all anomalies that occurred during 1999 can be attributed to La Niña. A La Niña (cold) event developed in March 1998 and extended into early 2000.*

one might not be able to forecast what *will* happen during a La Niña event, one can identify what is *less likely* to happen in some locations during La Niña. Here are some other examples:

- Northeast Brazil is less likely to have a drought
- Southern Brazil, Uruguay and Argentina are less likely to receive good rains for crop production
- Southern Africa is not likely to have a severe regional drought
- East Africa is not likely to have severe flooding
- Indonesia, Philippines, Papua New Guinea, and Malaysia are likely to have average to above-average rainfall
- Indonesia is less likely to suffer from uncontrollable tropical fires
- Central Chile is not likely suffer flooding (nor will flowers bloom in the Chilean desert)
- Strong, nutrient-rich coastal upwelling and arid conditions would re-appear along the Peruvian coast
- The Atlantic hurricane season will become more active
- India's monsoon is less likely to fail
- Precipitation in the southern part of the People's Republic of China is less likely to be excessive

While the mere mention of the possible onset of a La Niña can spark reactions from decisionmakers in various countries or corporations where those decisionmakers believe that there is a strong (usually adverse) La Niña impact (e.g., Philippines, Malaysia, Brunei, USA, Argentina), it is important to keep in mind that: (a) a sharp drop in warm SSTs in the tropical Pacific as happened in May 1998 does not assure that a strong La Niña will follow; (b) a La Niña does not always follow an El Niño; and (c) La Niña events vary in intensity, and that each level of intensity (weak, moderate, strong, very strong, and extraordinary) generates its own set of world-wide teleconnections.

# 6 The 1982–83 El Niño: a case of an anomalous anomaly

The *American Heritage* dictionary defines an anomaly as "a deviation from the normal order" and as "something unusual or irregular". El Niño is an anomaly in the sense that it is a deviation from *expected* changes in sea surface temperatures and in sea level pressure in the tropical Pacific. The 1982–83 El Niño can, therefore, be described as an anomalous anomaly. It was not recognized at the time it occurred in terms of either when or how people expected a "normal" El Niño event to develop. In addition to being unusual, it had reigned until early in 1998 as the most extreme El Niño event of the twentieth century. Now, the 1997–98 El Niño is viewed by scientists as having captured that title. In a 15-year period there have been two "El Niños of the twentieth century"! Nevertheless, the 1982–83 event, one that a United Nations (UN) report called an "abnormal natural phenomenon" holds a very special place in the development of scientific, political, and public interest in the phenomenon (UN ECLA, 1984).

The sea surface temperatures associated with the 1982–83 El Niño were well above normal in the central and eastern equatorial Pacific, reaching more than 4 degrees Celsius above average in some areas. Those concerned about natural disasters also considered it to have been a very extreme event, because of the numerous destructive climate-related impacts that took place in various regions around the world at the time. In the wake of this particular El Niño event, many world leaders and media, for the first time, were forced to pay more attention to the phenomenon. The 1982–83 El Niño also played a role in increasing the level of international support for developing a science program (called the Tropical Ocean–Global Atmosphere (or TOGA) program) to monitor, on a real-time basis, changes in the tropical Pacific along the equator.

As noted earlier, in 1982, several months prior to the recognition of the onset of the 1982–83 El Niño, two American researchers published an article about what they considered to be the "typical" El Niño. The "canonical" El Niño, as they called it. In a matter of about 6–8 months after the article was published (March 1982), the notion of a typical El Niño (one type) was challenged with the occurrence of an out-of-phase, and therefore unexpected (by almost all researchers) onset of the 1982–83 event. Edmund Harrison and Mark Cane (1984, p. 21) observed that

> The warm event in the Pacific in 1982–83 was unusual in many respects. Rather than exhibiting surface warming first along the northeastern [*sic*: northwestern] coast of South America in the spring, sea surface temperatures first significantly exceeded climatological [average] values along the equator in the eastern central Pacific during late summer.

In reviewing what had happened in 1982–83, Eugene Rasmusson and Phillip Arkin (1985, p. 183) wrote that the timing of the warming of water in the central Pacific was typical but that the warming along the Peruvian coast *followed* instead of preceded that warming. Anyhow, the warming of coastal waters occurred at the normal time of the year.

Thus the 1982–83 El Niño differed in both the timing and the location of onset from the set of post-World War II El Niño events that were used to identify the characteristics of a "typical" event. For example, anomalously warm sea surface temperatures appeared first in the central Pacific instead of off the coast of Peru. The warm sea surface temperatures moved eastward toward the South American coast, instead of first appearing along the coast and then moving in a westward direction away from it. It emerged later in the year (between June and August) than one would have expected (at that time) with a typical El Niño. The winds along the Peruvian coast did not weaken as expected when this El Niño began, even though a weakening of the westward-flowing winds was considered to be a necessary (but not sufficient) condition to spark its onset.

Another unexpected aspect of the 1982–83 event was the breadth and severity of its ecological and societal impacts. Because it began relatively late in the calendar year, societies affected by it were caught off guard, even those societies where the impacts related to El Niño were known to have a strong likelihood of occurrence. The fact of the matter is that, except in a few countries (Australia, Peru, Ecuador), an early warning in late 1982 that an El Niño was developing would not have evoked much activity to mitigate its potential impacts such as droughts, floods, fires or frost. Until 1983, El Niño was an obscure natural process. It was not viewed as having societal implications except for Peru, Ecuador, and Chile. Even in Peru, little attention (except perhaps in the fishing sector) was given to El Niño, as no reliable forecast system for identifying its onset had as yet been developed. The 1982–83 El Niño demonstrated to the scientific community, and to the federal agencies in the USA and Australia that funded El Niño physical science research at that time, just how much was yet to be learned about this potentially devastating natural phenomenon. It also demonstrated the potential value of long-range forecasts of El Niño's teleconnections.

By all published accounts, almost all scientists involved in El Niño research had failed to recognize the development of the 1982–83 El Niño for several months. Underscoring the scientific uncertainty surrounding this situation are the remarks of a senior El Niño researcher who had been in Peru in August 1982. As he left the country to return home, he announced that there would be no El Niño that year because the various early warning indicators that he had identified and relied on to forecast an event were not yet evident. Within the next few months at meetings of El Niño researchers in Miami, Florida and Princeton, New Jersey, scientists also concluded that there would be no El Niño event that year. As we now know with hindsight, within a matter of weeks their forecasts of "No El Niño" were proven wrong.

There were two notable warnings of the 1982–83 event. Australian meteorologist Neville Nicholls noted in mid 1982 (personal communication) that the Southern Oscillation Index was extremely low, a strong indicator of an impending El Niño. American meteorologist Eugene Rasmusson suggested in June that an El Niño was likely to emerge later that year, although no mention was made of its likely intensity. Once the El Niño was in full swing, the scientific community officially recognized that it had failed to identify the onset of the 1982–83 El Niño. A report summarized what had happened:

> The planning of TOGA was just getting under way in November 1982, when the strongest ENSO event thus far this century caught the scientific community by surprise. Unlike its predecessors during the previous three decades, the 1982–83 warm episode was not preceded by a prolonged "buildup phase" with strong trade winds along the equator, and it did not exhibit what had come to be viewed as the typical "onset phase" around April, characterized by El Niño conditions along the South American coast, which would later spread westward across the basin. In retrospect, it is apparent that the first indications of a major warming should have been evident in July and August 1982, when anomalous equatorial westerlies were observed to develop in the central Pacific, accompanied by strong sea level pressure rises at the western end of the Pacific basin. El Niño conditions did not become apparent along the South American coast until November, by which time the basin-wide warming had nearly reached its peak.
>
> (NRC, 1990, pp. 11–12)

In response to the undetected onset of the 1982–83 event, the scientific community implied that, like snowflakes, no two El Niño events were alike. This, however, appears to have been a knee-jerk response to having missed the forecast of this El Niño. The truth, of course, lies somewhere in between these opposing views – a typical El Niño or each one being different. Most likely there are types of El Niño event based on sets of characteristics, including, for example, where and when the sea surface temperatures begin to warm. Also, scientists still do not know how a global warming of the atmosphere over the next several decades, whether natural or human engendered, might affect the characteristics or behavior of El Niño or of its impacts. So, while each El Niño event provides researchers with more information about the phenomenon than they had before, they have come to realize that the El Niño puzzle is bigger than they thought. Forecasting its impacts around the globe is also more difficult than previously thought because those impacts are affected by what is going on at the same time with sea surface temperatures in other oceans.

## Impacts associated with the 1982–83 El Niño

Most, if not all, of the major weather anomalies around the world occurring in late 1982 and in 1983, especially droughts and floods in the tropics, were linked (more correctly, blamed) by one observer or another on this El Niño. Several El Niño articles, maps, and charts appeared at the time in the popular press suggesting the extent of the worldwide, continent-wide, national, and local impacts of the 1982–83 El Niño. Caution must be used, however, in attributing the cause of any particular climate-related anomaly or impact to a specific El Niño. Furthermore, the severity of societal impacts will vary according to the level of societal vulnerability to such extremes and not only to the intensity of a particular El Niño. Climate-related anomalies (e.g., droughts and floods) also result from a variety of local and regional conditions, even in the absence of El Niño events. Some examples of the alleged real and perceived societal impacts of the 1982–83 El Niño are shown in Figure 6.1.

The 1982–83 El Niño was also blamed for droughts in Sri Lanka, the Philippines, southern India, Mexico, and even Hawaii, along with severe, unseasonal and numerous typhoons in French Polynesia. It was credited with having played a role in suppressing hurricane activity along the Atlantic seaboard and in the Caribbean. In 1983, many of these events were reported to be record setting: the *worst* typhoon, the *most intense* rainfall, the *warmest* winter, the *longest* drought, the *fewest* hurricanes making landfall on the eastern USA, and so forth.

This El Niño was also associated with indirect societal and environmental effects. Indirect or secondary effects, however, are even more difficult than direct effects to attribute to an El Niño, as they could be the result of a variety of other causes. In 1982–83, these indirect effects took the form of dust storms in Australia and brush fires in the Côte d'Ivoire, Ghana, and Australia. In the USA, the 1982–83 event was blamed for adverse health effects such as encephalitis outbreaks in the East (the result of a warm, wet spring providing the proper environment for mosquitos), an increase in

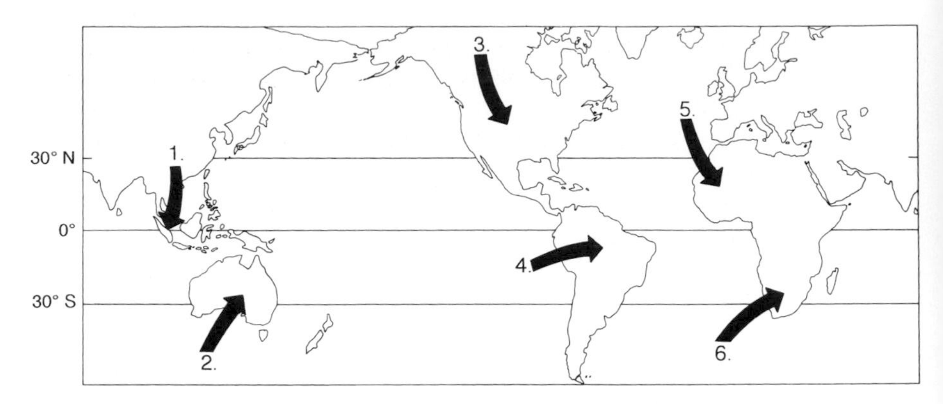


*Figure 6.1. Real and perceived impacts of the 1982–83 El Niño. According to Trenberth (1997), an El Niño occurred from April 1982 to July 1983.*

**Arrow 1.** *Indonesia was plagued with severe drought, resulting in reduced agricultural output, especially rice, and in famine, malnutrition, disease, and hundreds of deaths. This drought came at a bad time, as this country had made great strides toward self-sufficiency in food production. In the few years immediately preceding the 1982–83 El Niño, it was emerging as a rice exporter. This drought, however, coupled with worldwide recession, huge foreign debts, and declining oil revenues, had set back Indonesia's economic development goals for the near term.*

**Arrow 2**. *Australia had its worst drought this century. Agricultural and livestock losses, along with widespread brushfires mainly in the southeastern part of the country, resulted in billions of dollars in lost revenues. An Australian journalist wrote that "the drought is not just a rural catastrophe, it is a national disaster." The drought had been linked to El Niño.*

**Arrow 3**. *The eastern part of the United States was favorably affected by its warmest winter in 25 years. According to an estimate by the National Oceanic and Atmospheric Administration, energy savings were on the order of $500 million. (The opposite was the case, however, during the cold winter that accompanied the 1976–77 El Niño.) The United States once again was adversely affected by devastating coastal storms and mudslides along the California coast, flooding in the southern states, and drought in the north central states, reducing corn and soybean production. Salmon harvest along the Pacific northwest coast were sharply reduced.*

**Arrow 4**. *In addition to the highly publicized damage to infrastructure and agriculture in Peru and Ecuador as a result of heavy flooding during this El Niño, there were severe droughts in southern Peru and Bolivia. A* *major drought continued in northeast Brazil, adversely affecting food production, human health, and the environment, and prompted migration out of the region into already crowded cities along the coast and to the south. There also were destructive floods in southern Brazil, northern Argentina, and Paraguay.*

**Arrow 5**. *Large expanses in Africa had been affected by drought. For example, the West African Sahel, once again, had been plagued by a major drought. While the human and livestock deaths resulting from this drought appear to have been fewer than those in 1972–73, the food production situation was considered poor. The view that the Sahel had been in the midst of a long-term trend of below-average rainfall since 1968 gained credibility.*

**Arrow 6**. *Southern Africa witnessed some of its worst droughts in the twentieth century. In 1983, for example, the Republic of South Africa, a major grain producer in the region, was forced to import from the United States about 1.5 million tonnes of corn to replace what was lost in their drought. Zimbabwe, a regional supplier of food, also was devastated by drought and was forced to appeal for food assistance from the international community. Likewise, Botswana, Mozambique, Angola, Lesotho, and Zambia, and the so-called Black National Homelands in the Republic of South Africa had their economies devastated by the drought of 1982–83. This was not the case during the 1972–73 event.*

rattlesnake bites in Montana (hot, dry conditions at higher elevations caused mice to search for food and water at more densely populated lower elevations; the rattlesnakes followed the mice), a record increase in the number of bubonic plague cases in New Mexico (as a result of a cool, wet spring that created favorable conditions for flea-bearing rodents), and an increase in shark attacks off the coast of Oregon (because they followed the unseasonably warm sea temperatures). Even an increase in the incidence of spinal injuries along California's coast was blamed on El Niño (as a result of swimmers and surfers being unaware that the floor of the ocean along the coast had been changed as a result of the violent wave action that accompanied coastal storms). [NB: Today, beachwear for surfers carry a new brand name, El Niño (Figure 6.2), not only in the USA, but in Fiji as well!]

The map shown in Figure 6.3, based on information compiled by the US government's National Oceanic and Atmospheric Administration (NOAA), depicts the socioeconomic impacts associated with the 1982–83 event. A group in NOAA put the costs of the 1982–83 event at US$13 billion and an estimated death toll at 2100. Admittedly, these numbers were compiled rather quickly at the time, and the numbers might increase

*Figure 6.2. Graphic logo for a California beachwear company in 1997.*

sharply given a careful review of 1982–83 impacts. In time, the El Niño research community will improve its ability to sort out those distant worldwide climate-related anomalies that might correctly be blamed on an El Niño event from those impacts that have dubious linkages.

## El Niño's positive side

There has been an overwhelming tendency for people (myself included) to focus on the negative aspects of El Niño. However, with El Niño-related regional shifts in temperature and precipitation, one can expect that some regions as well as some human activities will benefit.

In 1588 José de Acosta published in Spain an account of his travels that included a visit to Peru. He reported on the coastal activities of the local populations, describing the balsas, the boats made of balsa tree logs bound together.

> Rowing up and downe with small reedes on either side, they goe a league or two into the sea to fish, carrying with them the cordes and nettes... They cast out their nettes, and do there remaine fishing the greatest parte of the day and night, untill they have filled up their measure with which they returne well satisfied. Truly it was delightfull to see them fish at Callao off of Lima, for they were many in number.
>
> (Acosta, 1588)

This interesting account suggests that there was a great abundance of fish relative to the level of technology available to fishermen at that time. To meet their needs from those marine resources, fishermen exploited the fish populations at levels that were well below the maximum levels of catch that the various fish populations could withstand and still survive. Contrary to popular belief, during El Niño events there are pockets of cold, nutrient-rich upwelled water that persist in the midst of a broad swath of warmer surface water, and those pockets are close to shore. One could

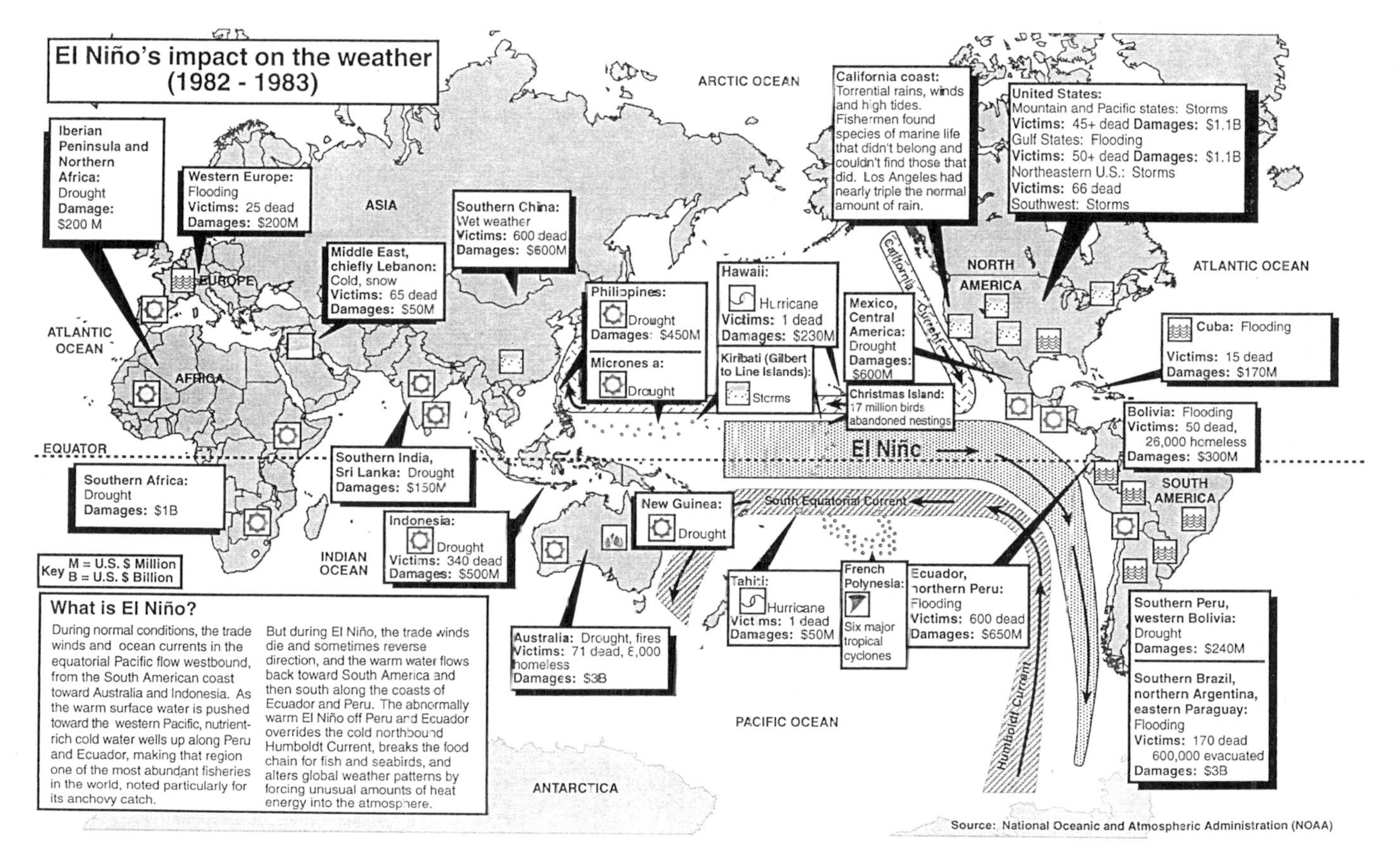


*Figure 6.3. The impacts and costs of the 1982–83 event that appear in the media are based on the information in this July 1983 map. National Oceanic and Atmospheric Administration's compilation of the damages of the 1982–83 El Niño. (Graphic courtesy of NOAA.)*

effectively argue that, during El Niño, fish in those cold water pockets would be easier to catch and with less effort of the fishermen. This would have been a boon to fishermen in ancient times, using their traditional technologies.

Today, the mention of El Niño suggests "bad times" for some species of fish (especially Peruvian anchoveta) and for the fishermen along the Peruvian and northern Chilean coasts. Yet, Acosta's remarks suggest that Peru's coastal waters were teeming with living marine resources hundreds of years ago. Is it just an El Niño that adversely affects fish availability and abundance? El Niño is not the only pressure on this living marine resource. There have been El Niño events for thousands of years and yet there has been an abundance of fish to catch. What has changed has been society's ability to capture fish in quantities that go much beyond just satisfying the local needs of humans for food. Today, if care is not taken in fishing using modern technology, the standing stock of adult fish could easily become decimated long before the fishermen have time to realize that they have destroyed the fish population they were exploiting.

Recently, George Philander (1998), commented on what he considered to be a basic misperception about El Niño, noting that local Peruvian fishermen had traditionally referred to El Niño episodes as "years of abundance". From their perspective, El Niño was a good thing. Thus some positive impacts of El Niño apparently occurred along the west coast of equatorial South America, e.g., the impacts of El Niño on phytoplankton, on the benthic community (marine resources that dwell near the ocean's floor), on increases in sardine, jack mackerel, shrimp, and scallop populations. American geologist Alfred Sears (1895), writing about Peru's coastal deserts in the 1890s, recorded several positive impacts of El Niño-related heavy rains in an otherwise perennially hostile arid environment.

In the early 1980s, German researcher Wolf Arntz also witnessed examples of some positive ecological changes that accompanied the 1982–83 El Niño:

> In many parts of the coastal desert, rains and unusually strong fogs [led] to a hitherto unknown outburst of vegetation that covered wide areas with carpets of flowers for several months, enabling local settlers to raise cattle, sheep, and goats ... Apparently, the seeds and bulbs of many plants survive in the desert for many decades until a strong El Niño creates appropriate conditions for the type of explosion observed this year.
>
> (Arntz, 1984, p. 37)

Another example of a positive effect of El Niño occurs off the coast of Ecuador. Under non-El Niño conditions, shrimp farmers buy shrimp larvae from hatcheries in the country, because the availability of wild shrimp off the coast is relatively low. However, with the appearance of El Niño's warm waters off Ecuador, there is a sharp increase in the abundance of wild shrimp and shrimp larvae, which can be captured for sale. The cost of wild shrimp larvae is lower than for those produced in hatcheries. This is financially good for the shrimp farmers, but that good news is tempered by the fact that many shrimp hatcheries operating during non-El Niño years go out of business (e.g., Cornejo-Grunauer, 1998). The sudden increase in the abundance of shrimp off the Peruvian coast is also a positive result of El Niño. For a while at least, some villages along Peru's coast fare well economically, as a result of an increase in the capture and export of shrimp (*Samudra Report*, 1999).

El Niño has been implicated in the below-average number of hurricanes and tropical storms in the Atlantic. These storms can be very devastating to countries in the Caribbean, around the Gulf of Mexico, and along the Atlantic seaboard. Atmospheric scientist William Gray (1993, p. 3) has noted that "the Atlantic basin ... experiences more seasonal variability of hurricane activity than occurs in any other global hurricane basin". According to Gerald Bell's and Michael Halpert's report on the 1997 hurricane season (June to November), over the North Atlantic, the suppressed activity during August–October 1997 was related to sustained high vertical wind shear across the Caribbean Sea and the western and central subtropical North Atlantic, and to the development of only one tropical storm from African easterly (westward flowing atmospheric) waves during August and September.

Their report also noted that, "overall, 9–10 tropical storms are observed over the North Atlantic in an average season, with 5–6 becoming hurricanes and 2–3 reaching intense hurricane status ... The suppressed 1997 hurricane season featured 7 named storms, with 3 of these systems becoming hurricanes and 1 reaching intense hurricane status" (Bell and Halpert, 1998, p. S28). According to Dr Chris Landsea (personal communication, 1999),

> the net tropical cyclone activity was only 52% of normal for the season as a whole. El Niño's impact on tropical storm activity was most evident during August to October, when only three systems developed. In fact, a record low (since the beginning of the aircraft reconnaissance era in 1944) of only one system formed during August and September, normally the months of peak tropical storm activity. These conditions contrast with the previous two very active Atlantic hurricane seasons which featured 19 named storms in 1995, 11 of which became hurricanes, and 13 named storms in 1996, 9 of which became hurricanes. During these two years the "net tropical cyclone activity" was 229% and 198% of normal, respectively. These years were La Niña years and during La Niña events the number

of hurricanes increases sharply when compared to the number that occurs in El Niño years.

## Limits of utility of the 1982–83 (or any other single event) case study

Today, just about everyone who knows anything at all about El Niño refers back to the 1982–83 event in much the same way that recent droughts in North America's Great Plains spark comparisons with those of the Dust Bowl Days in the mid-1930s. The reasoning is as follows: because it was the biggest El Niño, its signal and some of its impacts were highly identifiable. In science jargon, scientists would say that the El Niño *signal* surpassed the *noise* surrounding it. This is not often the case during weak or moderate El Niño events, where regional or local climatic and environmental processes can overshadow the influence of a distant El Niño. In areas where the public has become aware of El Niño, people tend to expect that future El Niños are likely to produce similar teleconnections in the same locations where they occurred during 1982–83. In fact, it was the suggestions by scientists to the media that the 1997–98 event would be like the 1982–83 event that generated a high level of public interest in that El Niño.

Some anomalous climate events, such as drought in northeast Brazil, flooding along the northern Peruvian and southern Ecuador coasts or in southern Brazil, or drought in eastern Australia and in Indonesia, have relatively strong and reliable linkages with El Niño. However, many others, such as torrential rains and flooding in southern California, bubonic plague outbreaks in the southwestern USA, or widespread uncontrolled forest fires in Malaysia or Indonesia, are not to be expected during each and every future El Niño. Thus, while the 1982–83 El Niño is a very informative, instructive and useful case study, lessons drawn from it must be used with great care. It is only one case for study, albeit an important one.

It is important to emphasize that the actual level of impact of an El Niño depends on the local and regional climatic and oceanic conditions that exist at the time of its onset, as well as on its intensity. Even if two El Niño events were to have the same physical characteristics (e.g., timing, intensity), their consequences could vary greatly from one place to another, and could even vary in the same location at different times, depending on a host of societal as well as environmental factors at the time that they occur. For example, southern Africa is usually drought prone during El Niño events, as was the case in 1982–83 and in 1991–92. However, during the intense 1997–98 event, the region had sufficient precipitation and water resources for agricultural production and national food needs. This unexpected wetness during a major El Niño was, in large measure, because of the unusual increase of sea surface temperatures in the western part of the Indian Ocean, which happened to coincide with El Niño.

> El Niño is but one of many influences upon the global atmosphere. During any particular event, the other processes may reinforce or obliterate the distant influence of El Niño. Climate anomalies observed in other ocean basins during El Niño can result from anomalous surface wind in those regions. The wind anomalies may or may not be "caused" by the El Niño.
>
> (Hansen, 1990, p. 6)

Until the onset of the 1997–98 event, that in 1982–83 had been the most analyzed El Niño, having been reviewed by atmospheric scientists, ecologists, hydrologists, fisheries biologists, ornithologists, oceanographers, social scientists, and economists, among others. Perhaps the overriding major benefit to the *scientific* community of the devastating 1982–83 event was that it accelerated the planning and development of the major research and monitoring effort called TOGA. Although some workshops were held in the mid 1980s in Ecuador, Peru, Chile, Australia, South Africa, and elsewhere to determine the regional impacts of the 1982–83 El Niño, few research activities, however, focused on its societal impacts.

As important as the 1982–83 event was in catalyzing interest in and financial support for El Niño research, it has become overused as an example of typical El Niño-related devastation. The reason is that, in the past 15 years or so, three or more El Niño events have taken place: in 1986–87, one (or more) in the 1991–95 period, and one in 1997–98. It was not until the last El Niño of the twentieth century – in 1997–98 – that an effort was undertaken to make an objective accounting of the societal and economic impacts and consequences of those events (e.g., see Sponberg, 1999).

## The forgotten 1972–73 El Niño

The 1982–83 El Niño overshadowed all the El Niño events that had preceded it, just as the 1997–98 event has overshadowed the 1982–83 El Niño. However, as important as these or earlier events may have been at the time, it was actually the 1972–73 El Niño that caused a steplike increase in scientific interest by enticing scientific researchers to pay closer attention to the phenomenon and to its role in the global climate system.

The 1972–73 El Niño event was considered at the time as "certainly the

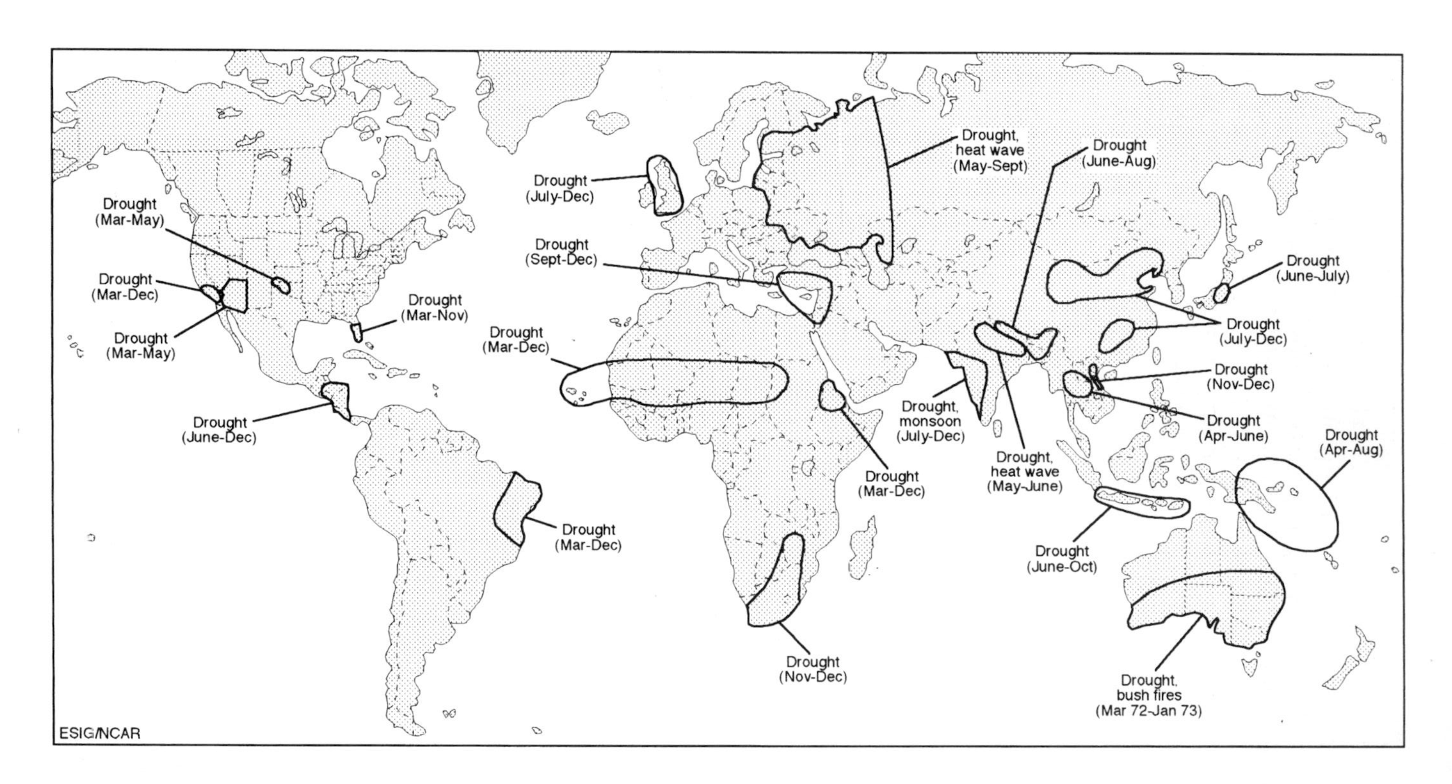


*Figure 6.4. Drought conditions worldwide in the period of March–December 1972. According to Trenberth (1997) an El Niño occurred from April 1972 to March 1973.*

most intensely observed since 1891" (e.g., Cushing, 1982, p. 290). As noted earlier, the political and social setting for this event help to highlight its potential importance to science and society. For example, in the early 1970s, for the first time since the end of World War II, global food production declined. Anomalous climate in the early 1970s sparked droughts, which some observers suggested were linked to El Niño, in the Soviet Union, West Africa, Ethiopia, India, southern Africa, Australia, Central America, Brazil, and Indonesia. Global fish landings in total also declined for the first time since the end of World War II. The adverse impacts on agriculture and fisheries of climate anomalies in the early 1970s served as the catalyst for convening the first World Food Conference in Rome in 1974. This was followed throughout the decade by a series of UN conferences on other global issues, such as population (held in Romania), human settlements (in Mexico), water (in Argentina), desertification (in Kenya), climate (in Switzerland), and technology (in Austria).

It was the 1972–73 event that brought the El Niño topic to the forefront of the scientific research agenda. The worldwide costs in lives lost and properties damaged as a result of the 1972–73 El Niño event have never been calculated. However, a qualitative picture of the drought damages associated with that El Niño event might be suggested by the impacts map shown in Figure 6.4.

The climate anomalies of 1972 sparked a resurgent interest in the study of climate (climatology) and precipitated the development of a subfield of multidisciplinary research that has since become known as climate-related impacts. Such research assessments have focused on the interplay of climate variability on managed and unmanaged ecosystems and human activities. Thus the 1972–73 event clearly merits a place in an El Niño "Hall of Fame," as *the* event that energized the oceanographic, atmospheric, and biological research communities, and also prompted some of the first papers on the societal impacts of El Niño.

## El Niño Olympics

There have been at least two meetings that I know of which had the title, "Is This [1997–98] the El Niño of the Century?" One was held in Peru in November 1997 on the proverbial eve of El Niño's impacts in that country. The other was held in April 1998 in Southern California following El Niño's impacts in North America (more specifically, in California). The media, and even policymakers, have many times over referred to this El Niño as the twentieth century's record-breaker.

The major El Niño of 1997–98 has surpassed the *first* "El Niño of the century" (1982–83) in both the magnitude of warming and in the cost of its impacts worldwide. Some have referred to it as the *second* "El Niño of the century". In the last days of the twentieth century it became apparent that the 1997–98 event had captured the title.

Researchers are in a position, as evaluators of El Niño who hold the public's trust, to say anything they wish to say about El Niño or their forecasts of it. Much of the public will accept what they hear at face value. We must then approach with great care the question of whether to label an El Niño event as the most intense one in a century.

Which El Niño was *the* event of the past century is actually a useful scientific question. But such a question raises other concerns of equal or greater importance. What, for example, should be measured in order to determine whether the 1997–98 El Niño merits such a title? From the standpoint of the public, people will consider this to have been the "winner" of the El Niño "Olympics," if the cost of the devastation associated with it in areas with which they happen to be familiar had been greater than in other El Niño years. Instead, they may choose to believe whatever the media report or whatever a group of scientists reports to the media. The reality, however, may be something that El Niño researchers do not want to hear: that El Niño is a natural phenomenon that they do not yet understand well enough to answer this question.

How, then, should scientists measure this event? By focusing on the degree of above-average sea surface temperatures in the central Pacific? By relying on how anomalously warm the water becomes off the coast of Peru? By focusing on how widespread its global impacts have been? Should they rely on the costs of the devastation associated with El Niño? Or, should they rely on the extent of media coverage, political interest, or public awareness of an El Niño to decide which one was the event of the century?

In fact, several of the El Niño events in the twentieth century merit the distinction of the El Niño of the century, but for different reasons. For example, the 1925 El Niño was devastating for Peruvian communities and ecosystems (e.g., Murphy, 1926). Although it was a little understood phenomenon at the time, even in Peru, its impacts were notable enough to capture the attention of the Peruvians. In that year Peruvians began to collect rainfall information in a systematic way in response to torrential rains and flooding in their northern regions.

The 1939–41 El Niño was a more controversial one. Until recently, it was considered to be the longest one of the twentieth century, running across 3 years. What seemed to be a continuous El Niño between 1991 and 1995 (according to some researchers) casts a shadow on that designation. The 1957–58 El Niño could also be considered for an "Olympic" title because it was observed accidentally (i.e., it was not part of any planned scientific experiment) during the International Geophysical Year taking place at that time. It sparked the interest of a small number of key scientists.

The 1972–73 event was a big one, although not the biggest, even up to that time. Its connection to the collapse of Peru's fishing industry captured the attention of the international press and of some physical, biological, and social science researchers. This El Niño could earn the title of "El Niño of the century," because it started researchers on the proverbial "slippery slope" of interest in the phenomenon.

Even the relatively mild El Niño in 1976 could seek to capture the title as well, but for reasons that are not so obvious. Researchers seem to agree that there was a shift in the behavior of El Niño events in the mid-1970s, following this event. (It is important to note that some researchers have referred to this as the 1976–77 El Niño, while others have suggested removing it from the list of legitimate El Niño events because it did not meet their criteria). Since the mid-1970s there have been about twice as many El Niño as La Niña episodes. This was different from its behavior during the previous few decades when La Niña events seemed to dominate.

Until recently, the 1982–83 event firmly held the title as the "El Niño of the century". As noted earlier, it was extraordinary in its size, unexpected in its timing, and devastating in its impacts. It captured the attention of the scientific community and the funding agencies. This El Niño sparked public awareness to such an extent that such popular magazines as *Reader's Digest* (December 1983) and *National Geographic* (February 1984), with readerships measured in the millions produced articles about it. It led to the development of the monitoring system across the equatorial Pacific known as the TOGA-Tropical Atmosphere Ocean (TAO) Array, using satellites and on-the-spot (*in situ*) measurements from fixed and drifting ocean buoys. The event was the standard against which successive El Niño events have been measured.

The 1986–87 event deserves special consideration, but not for its size, timing, or impacts. It was the first El Niño to have been forecast to the public by researchers. Their *public* prediction went against the existing (at that time) unwritten forecasters' code not to go to the media with such experimental climate projections. Despite the flak they received at the time from the scientific community for "going public," it was only a few years later that all groups involved in developing forecasts of El Niño's onset were going to the media with their projections.

The 1991–92 El Niño started out as a typical event, spawning typical El Niño-related impacts around the globe; droughts and floods occurred in locations in which they might have been expected (e.g., Australia, Indonesia, and southern Africa suffered major droughts, while northern Peru and southern Ecuador suffered from excessive rains and flooding). But this El Niño did not go away. Late in 1992 it appeared to be in its decay phase, only to re-emerge in 1993 as another El Niño. It did this again in 1994. To the Australians, it was an event that caused a 5-year drought in their country. To Peruvian fishermen, it appeared to be three successive, weak El Niño events with little impact on their highly productive coastal fishing operations. This seemingly multi-year El Niño prompted the public (at the instigation of some researchers) to blame global warming for such a long El Niño. The early 1990s event(s) is (are) claimed by some observers to have been the longest of this century, making it a contender for the title of "El Niño of the century". Two researchers (Trenberth and Hoar, 1996) gave the 1991–95 El Niño event(s) a probability of occurrence of once in 2000 years, a probability that has been challenged by other researchers.

The latest El Niño of 1997–98, the last warm event of the twentieth century, has manifested unanticipated characteristics: it developed earlier than expected, stayed strong longer than expected, grew bigger than expected, and was hotter than expected. On the basis of these characteristics alone, it is a strong contender for the title. But that is not all it did. It prompted the biggest media blitz of all previous events, capturing and sustaining the interest of most national policymakers worldwide. It sparked the organization of numerous bureaucratic units in various governments to deal with the phenomenon and its impacts, especially in the form of "Task Forces". It has been the most observed El Niño ever, from onset to decay. Some scientists, policymakers, and members of the media have suggested (yet to be proven) that its impacts on societies and ecosystems have well surpassed those attributed to the previous El Niño champion, the 1982–83 event. In essence, just about everything that happened during this El Niño (climate-related or not) has been blamed on it. It prompted the development of numerous websites on the Internet, in a variety of languages devoted to all kinds of aspects of El Niño.

Regardless of indicators used by the various judges, it is most likely that the 1997–98 El Niño will go down as the "event of the twentieth century," in the public's perception at least. Most people do not recall the impacts of events before 1983, nor do they remember much about the 1982–83 event. In addition, people tend to weigh recent events more heavily (more importantly) than those in the distant past (i.e., they discount the past).

As a lone judge, however, I object to this search for the "event of the century" and am throwing away my score card. Each event has its own unique qualities. Each contributed to our understanding of the El Niño process. In a way, this issue has the ring to it of a parent answering the question, "Which one of your children do you like best?". The answer, to the public at least, is likely to be, "I like all of them equally well". The parent may be answering truthfully, but most likely he or she is measuring each one by a different set of criteria. The same can be said for evaluating El Niño events. However, it is important to make the conditions (and indicators) by which we make our comparative assessments known. Before anyone ranks El Niño events, he or she must make the criteria explicit.

# 7 Forecasting El Niño

Different words are used to describe predictions of what might happen in the future. A popular word for this is forecasting. Another might be projection – yet others, scenario, outlook, or prognostication. Sir Gilbert Walker used the term "foreshadow". Each of these words is used in order to present a possible image of some aspect of the future. These words are found in the scientific literature, especially in El Niño research, and in the popular media as well. Some scientists insist that they are not making forecasts but do acknowledge that they are offering projections. To a person on the street, this semantic distinction has little meaning: "A rose by any other name is still a rose," and a forecast by any other name is still a forecast!

> fore·cast \ fō(ə)r-käst \ *vb* **forecast** *also* **fore·cast·ed; fore·cast·ing** *vt* (15c) **1 a:** to calculate or predict (some future event or condition) usu. as a result of study and analysis of available pertinent data; *esp*: to predict (weather conditions) on the basis of correlated meteorological observations **b:** to indicate as likely to occur **2:** to serve as a forecast of: PRESAGE ⟨such events may ~ peace⟩ ~ *vi*: to calculate future *syn* see FORETELL – **fore·cast·able** *adj* – **fore·cast·er** *n*
>
> *Merriam Webster's Collegiate Dictionary*, 10th edn, 1993

Most recently, an anonymous reviewer of one of my El Niño articles critically objected to my references to El Niño forecasts, suggesting that "forecasters do not predict 'El Niño'. They forecast tropical sea surface temperatures or they forecast the teleconnections it implies." Yet titles of scientific publications and the headlines in the popular media on El Niño forecasts show that researchers as well as operational forecasters do attempt to predict El Niño, in the sense that they expect their forecasts or projections to come true.

El Niño modelers have been producing what they call "experimental" forecasts. These are forecasts (or projections), based on computer or statistical output, about the state of sea surface temperatures in the central Pacific. They put those forecasts in a report issued quarterly on the Internet for all to see. To the public these are in essence forecasts of what is to come. For their part, some researchers who publish their model's projections in such a way claim that they are experimental and not really for public use. They do not say that they are not *ready* for public use. However, several of these researchers share their projections with the media on a continuing basis. Almost immediately the adjective "experimental" disappears, turning these statements into El Niño-related forecasts as far as the media and the general public are concerned. The bottom line is that all descriptions attempt to provide to a target audience (the public, policymakers, or other scientists) a glimpse of what specific forecasters think the future climate might be like.

Forecasts are just probability statements about what might occur. They have a chance of being right, but they also have a chance of being wrong. Even though forecasts might prove to be right some of the time, one should also expect them to be wrong sometimes as well. With regard to the forecasts of the 1997–98 El Niño, various groups claim to have successfully forecast its onset. Even if this were true (which I do not believe was the case), some of those groups had a poor record of El Niño or La Niña forecasts in previous years. For example, one California researcher claiming success in forecasting the 1997–98 event had also forecast a strong record-breaking El Niño in the preceding year (1996), which turned out to have been a major La Niña event. Forecasting El Niño has become a rewarding pastime from the standpoint of research funding. The public, and especially the media, must scrutinize more carefully claims of forecast success.

Attempts to assess the value to society of a forecast system must not be based on the success or failure of any single forecast. Value can best be measured by objectively assessing the success or failure over time of a series of forecasts. El Niño- or Southern Oscillation-related forecasts should be assessed in this way. Forecasts must be used with care. The more dependent we become on them, the more care we must take with their use. El Niño forecasts do not come with guarantees; there are identifiable risks (as well as benefits) associated with their use.

## The value of a forecast: what is versus what ought to be

The following is the opening paragraph in a Queensland (Australia) government booklet on El Niño and the Southern Oscillation.

> "If I'd known that it was going to stay so dry, I would have sold more cattle early in the season," said the grazier. "If I'd known it was going to be so wet, I wouldn't have bought that new irrigation pump," said the farmer. If those whose livelihoods depend on the weather had a better idea of the coming season, they could make better decisions.
>
> (Partridge, 1991, p. 1)

This paragraph expresses succinctly the perception, if not the core belief, of most people that having more information about future weather or climate conditions has to be better than having less. Assessing the value to a farmer, to a grazier, or to a society is not such a simple, straightforward task. As there is no such thing as a perfect weather forecast, forecasts in Queensland are issued in terms of probability. The Australian government booklet, *Will It Rain?* noted that "We can improve the probabilities based on historical rainfall records. Meteorologists have found that the Southern Oscillation can allow better prediction of rainfall in the coming season." The report continued,

> Monthly rainfall totals can be checked mathematically to show the distribution of wet or dry years and, from this, we can express the chances, or probability, of getting more than a certain amount of rain in any month. But a 66% probability of getting more than, say, 100 mm of rain still means that one year in three we could make a wrong decision.
>
> (Partridge, 1991, p. 1)

The potential value of reliable predictions of El Niño was recognized by scientists at least a few decades ago. In 1961 Jacob Bjerknes referred to the potential value to fishermen of forecasting El Niño's onset, suggesting that

> This particular meteorological phenomenon will soon be photographically reported from "Tiros" or (later) "Nimbus" type weather satellites. With such data at hand, it may become possible to discover the first beginnings of a developing "El Niño" at sea early enough to issue useful "El Niño" warnings for the coastal fisheries.
>
> (Bjerknes, 1961, p. 219)

Later, University of Hawaii oceanography professor Klaus Wyrtki and his colleagues also commented on this possibility:

> The economy of Peru is strongly influenced by its fishery, as is the world market of protein for animal feed, and so a prediction of the occurrence of El Niño would be a valuable guide for long-range economic planning... [A] capability to predict this event would contribute to the understanding of the large changes occurring in our weather and climate.
>
> (Wyrtki *et al.*, 1976, p. 343)

However, what ought to be the value in the best of all possible worlds may not be what is actually achievable, given the various social, economic, cultural, and political constraints in a society affecting the ideal use of such information at any given point in time. These are the two different perspectives on a forecast's actual value to society.

If you were to ask a fisherman, farmer, or an industry representative what one might do with perfect knowledge about a future weather event, he or she would likely respond with numerous tactical uses of that information. For example, with regard to Peru's anchoveta fishery, some might argue that the information in the forecast could be used to protect the fishery against overexploitation: tie up the fishing boats during El Niño so that they would not take too many fish from the sea at a time when the fish population is most vulnerable; shorten or close the fishing season; limit the fish catches; change the mesh size of the fishing nets so anchoveta can escape capture; and so forth.

However, from the perspective of the fishing sector, many of the tactical measures to protect the anchoveta population in the long run would place major financial hardships on the fishermen, the boat owners, the crews, ancillary industries such as shipbuilding, and banks that loaned funds to the fishing industry. If fishermen are not allowed to fish, how can those who are dependent on fishing for a living cover their monthly payments to the bank for the boats or equipment? The reality is that fishermen continue to demand the right to fish, even during El Niño. The only thing between them and their all-out exploitation of the fish are government decisions to stop all fishing efforts for months at a time (closed fishing seasons are called *vedas* in Spanish). Fishmeal processing plants, too, have a constant need for fish catches to process into fishmeal and oil. Demands in the international marketplace for Peru's fishmeal exports also put pressure on the Peruvian fishing sector to keep on fishing, especially when fishmeal prices are high. Some argue that "we might as well catch the fish before El Niño's high temperatures kill them off".

And then there are equity issues. Once a forecast is decided upon, who gets the forecast first? How a forecast is delivered and to whom, as well as when, determines who has an advantage over others who do not receive the forecast at all or receive it later. While forecasters may consider putting their forecasts on the Internet for all to see at the same time as treating all forecast users equally, fishing companies will probably have immediate and direct access to the Internet and information on real-time changes in oceanic conditions. However, poorer artisanal fishermen will have no such access to fishing (or to El Niño) information. Whose responsibility is it to assure equity with regard to forecast dissemination? These questions are now being raised by multidisciplinary researchers concerned with forecast application issues (see Broad, 1999; Pfaff *et al.*, 1999).

Thus, when the public listens to scientists as they speculate about how an El Niño forecast might be of value months in advance of its onset, it is important to keep in mind that it will be difficult to realize all the potential that they might identify in theory.

## Going public

A scientific prize will probably go to the El Niño group or researcher who first develops a way to produce highly reliable forecasts of the *onset* of El Niño several months in advance. Nevertheless many researchers are interested mainly in understanding other pieces of the El Niño puzzle, focusing, for example, on the oceanographic or atmospheric or biological processes associated with an event.

Some researchers base their El Niño forecasts on changes in the ocean environment, such as changes in sea surface temperatures or changes in ocean currents. Others base their projections on observed changes in the atmosphere, such as changes in sea level pressure differences across the Pacific basin or surface wind speed and direction. Still others use statistical or analogy methods or computer models to forecast the onset of events.

The scientific community, however, is not yet certain that all of the puzzle's pieces are as yet on the table, since new aspects of El Niño processes have been uncovered with each successive event. Until the mid-1980s, scientists were reluctant to "go public" with their thoughts, let alone their forecasts, about whether an El Niño event may be in the offing. While many researchers may still feel uncomfortable about making such public pronouncements, others now feel compelled to speak out about their interpretations of the existing conditions in the tropical Pacific Ocean. If they do not do it, others will.

As confidence increases in different forecast methods, more and more scientific groups are going public with their projections about the onset of El Niño events, using whatever indicators they feel are most relevant, and using whatever medium they feel best suits their goals – the print or broadcast media or the Internet. Their forecasts compete with those that are officially produced by, say, the US National Centers for Environmental Prediction (NCEP) or by the Australia Bureau of Meteorology (BOM).

Does the present state of knowledge about forecasting the onset of an El Niño merit the marketing of forecasts to the public as a "done deal"? Are the various El Niño forecasts usable by society, even though their reliability over the long term has yet to be determined? El Niño forecasters do believe that the reliability of El Niño forecasts has already been proven. There are some good examples of forecasts that have proven to be correct. There have also been missed forecasts. Both kinds of example are presented in the following sections.

## Forecast successes

### *USA: 1986–87*

The first successful public forecast was issued by Mark Cane and Stephen Zebiak of the Lamont–Doherty Earth Observatory at Columbia University in the early months of 1986. These researchers decided to go public with their forecast of an impending El Niño for the 1986–87 period.

**re•li•abil•i•ty** \ ri-ˈlī-ə-bi-lə-tē \ *n* (1816) **1:** the quality or state of being reliable **2:** the extent to which an experiment, test, or measuring procedure yields the same results on repeated trials
**re•li•able** \ ri-ˈlī-ə-bəl \ *adj* (1569) **1:** suitable or fit to be relied on: DEPENDABLE **2:** giving the same result on successive trials – **re•li•able•ness** *n* – **re•li•ably** \ -blē \ *adv*

*Merriam Webster's Collegiate Dictionary*, 10th edn, 1993

Their forecast was based on their simple oceanic and atmospheric model (called a coupled model) focused on the equatorial Pacific region (Cane *et al.*, 1986). Many of their scientific colleagues expressed concern because of Cane and Zebiak's decision to present their forecast to the public, believing that their research findings were not yet sufficiently reliable for public consumption. The development of a moderate El Niño in late 1986, however, proved that their projection was correct. Soon, other researchers began to issue El Niño forecasts to the public.

*Ethiopia: 1987*

Following a major famine in Ethiopia in the mid-1980s, when several hundreds of thousands of Ethiopians perished, the attention of international political leaders focused on the urgent need for a famine prevention program for the Horn of Africa. One activity that resulted was the regional development of early warning systems for food security and famine prevention. Another activity was to search for early indicators of changes in rainfall, which have been closely associated with crop failures. In this connection, Ethiopia's National Meteorological Services Agency (NMSA) started to issue seasonal forecasts and brought El Niño considerations to the attention of high-level authorities concerned about the possibility of drought during the 1987 main rainy season (called *kiremt*) from June to September. Therefore, the NMSA was well aware at that time of the apparent statistical linkage of El Niño thousands of kilometers away in the equatorial Pacific to the increased potential for failure of the rains during Ethiopia's main rainy season. One particular report, entitled *The Impact of El Niño on Ethiopian Weather*, issued in December 1987 (the end of an El Niño year), noted that

> several investigations have revealed that the rain-producing components in Africa are either weakened or displaced or both in ENSO years . . . However, it is not only the ENSO events that cause the recurrence of droughts. Other investigations have indicated that the warm sea surface temperature anomalies over the southern Atlantic and Indian oceans

ETHIOPIA

ANALOGUE BASIS FOR FORECAST

| | "Small Rains" | "Big Rains" |
|---|---|---|
| I. ENSO | Heavy | Light |
| II. Post ENSO | Light | Heavy |
| III. Anti-ENSO | Light | Heavy |
| IV. Normal | Variable | Stable |

*Figure 7.1. Chart produced by Ethiopia's National Meteorological Services Agency in 1986–87 that suggested the possible impacts of El Niño–Southern Oscillation (ENSO) on its "Small Rains"* (belg) *season (mid-February to mid-May) and "Big Rains"* (kiremt) *season (June to September). Anti-ENSO refers to a La Niña (cold) event.*

> during the rainy season have considerable influence on the recurrence of droughts in Africa . . . The better understanding of the occurrences of ENSO events and their influence on Ethiopian weather are indispensable aids in preparing and disseminating long-range weather prediction.
> (NMSA, 1987, pp. 1–2)

By looking at Ethiopia's yearly climate records for both El Niño and non-El Niño years and identifying what their seasonal rainfall patterns had been like during those anomalous periods, Ethiopian meteorologists were able to suggest with some degree of confidence to government officials what the prospects might be for adverse changes in rainfall, and, therefore, for reduced agricultural production, in the upcoming seasons. This analog basis for making such forecasts (Figure 7.1) was a popular method among North Americans and Europeans in the middle decades of the twentieth century. Today, however, it is not considered to be a robust (i.e., reliable) approach to seasonal forecasting, unless it is used in combination with other information.

Whenever a forecast of El Niño is issued (by just about anyone), someone is likely to take action based on it. The forecast of the onset of El Niño for late 1986 and 1987 was no different. It generated considerable concern about a possible return of famine conditions to Ethiopia. Ethiopian meteorologists and government took the prospects of an ENSO warm event very seriously and sought to modify the "normal" behavior of farmers. The government encouraged farmers to engage in all-out production during the "Small Rains" (the short rainy season) that occur from mid-February to mid-May in anticipation of increased losses that could

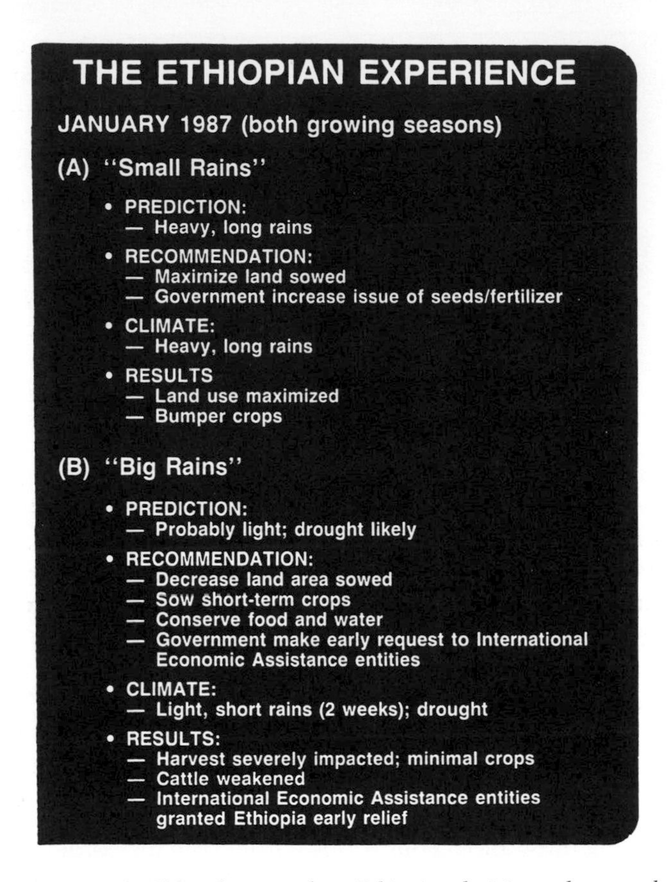


*Figure 7.2. Chart depicting how Ethiopian decisionmakers used the forecast of El Niño in 1986 to maintain national food security in 1986–87.*

result from severe drought during the main rainy season. As a result, this El Niño forecast enabled the Ethiopian government and farmers to take mitigative action (Figure 7.2). The anticipated severe meteorological drought did indeed occur during the season of the "Big Rains," but the government's action reduced the amount of food relief that would otherwise have been needed from the international community. Above all, because of the timely action taken by the government, no human life was lost in the region.

### *Cane/Zebiak and the National Meteorological Center: 1991*

A few years later, in 1990, researchers at various research centers decided to issue their forecasts of El Niño for that year to the public. Cane and Zebiak's model, however, did not agree that an El Niño would develop. An El Niño event failed to materialize in 1990, but in the early months of 1991, Cane and Zebiak, using the relatively simple wind-driven ocean model, forecast the onset of an El Niño for late 1991. Once again, they were correct. Forecasters at the US Government's National Meteorological Center, using a statistical model, also correctly forecast an El Niño. With these two successes to point to, the forecast community became much more encouraged about the prospects of developing a truly usable reliable long-range El Niño forecast. The Cane–Zebiak model became the one for other modeling groups to "beat," as this model had become the leading (so-called flagship) forecast model.

### *Northeast Brazil (the Nordeste): 1991*

Drought in northeast Brazil is not an uncommon occurrence. In fact, the Brazilian parliament has held debates at least as early as the mid-1800s about how to mitigate the impact of drought in this vulnerable region. Many famous novels have been written about the plight of inhabitants of the region since then, the most famous of which is probably Euclides da Cunha's *Rebellion in the Backlands* (Cunha, 1944).

Northeast Brazil provides another example of a successful El Niño forecast. The governor of the state of Ceará had become convinced by Brazilian and international scientists that El Niño events were associated with recurring severe droughts in the Nordeste region. A climate monitoring bulletin of the Meteorological Foundation of Ceará (FUNCEME) mentioned the linkages as follows:

> The El Niño phenomenon begins to affect the coast of Peru. When this localization of El Niño is finished, its interference in the rainy season [in the Nordeste] takes effect. During March, when El Niño becomes established on the northwest coast of South America, the regular rainfall in the entire Northeast region becomes reduced.
>
> (FUNCEME, 1992, p. 13)

The state of Ceará supports FUNCEME, both financially and morally, in its attempts to forecast the probability of regional droughts. In fact, FUNCEME had been created to inform government leaders about the regional climate situation from one season to the next and from one year to the next.

FUNCEME's major success occurred in December 1991, when its forecasters issued a warning a few months in advance of severe drought, based on an assessment from the US National Weather Service of the possibility of the onset of an El Niño. The governor, believing the forecasters, traveled throughout Ceará to encourage people to respond to the threat of drought by planting crops that could grow and mature in a

Table 7.1. *Comparison of 1987 and 1992 drought impacts in Ceará*

| Year | Precipitation (% of average) | Grain production (tonnes) | Grain production (% of average) |
|---|---|---|---|
| 1987 | 70 | 100 000 | 15 |
| 1992 | 73 | 530 000 | 82 |

drier, shorter-than-usual growing season. In addition, warnings were issued to the people in Fortaleza, the capital city of Ceará, that there would be severe urban water shortages in the event of drought. As a direct response to this warning, rationing was imposed by the government as a precautionary measure. The governor also decided to support the construction of a new dam to "stockpile" water resources.

With a drought in Ceará in 1992, these pro-active responses were considered to have been successful, as can be seen by comparing the severity and impacts of the 1992 drought with a similar one in 1987 (Table 7.1).

The assumption of two authors (Lagos and Buizer, 1992) was that the major difference between the impacts on agriculture of these two El Niño-related droughts was the availability and use of an El Niño forecast for 1991–92. El Niño forecasts apparently influenced the crop production activities of farmers.

### *Australian examples*

The high variability of rainfall in northern and eastern Australia has been strongly associated with the influence of the El Niño phenomenon and especially with the Southern Oscillation. The 1982–83 El Niño event visited upon Australia its worst drought on record up to that time. Its most recent prolonged drought in the early 1990s was also costly, with adverse impacts estimated at more than AUS$3 billion. Although droughts have plagued Australia in the past and their connections to variations in sea level pressure across the Pacific basin had been known for decades, it was the negative impacts of the bush fires, dust storms, agricultural and livestock losses, and so forth in 1982–83 that prompted interest in El Niño among the Australian public, policymakers, and researchers.

The Southern Oscillation Index (SOI) has been used in combination with other relevant information to forecast the possible occurrence in parts of Australia of good and poor rains. With at least some lead time, decisionmakers would have options to manage the exploitation of their climate-dependent resources (such as their grazing lands) more efficiently. Graeme Hammer, Principal Scientist in Queensland, has noted that

> there is potential value of seasonal forecasting to government in considering a range of policy issues (macroeconomics, trade, taxation) as a consequence of the pervasive influence of ENSO on the Australia economy.
>
> (Hammer, 1995, p. 6)

Hammer, along with colleagues such as Roger Stone, has been quite active in applying El Niño information to the needs of farmers and graziers at the local level. Several Australian studies have been undertaken on the application of El Niño and of Southern Oscillation information in decisionmaking. The studies have focused on possible linkages between ENSO and changes in the frequency, location, and intensity of tropical cyclone activity in the eastern Australian region, the SOI and low river flow in the Darling River, droughts and El Niño events, El Niño events and fluctuations in waterfowl numbers in southeast Australia, trends in the SOI and wheat and sorghum yields, and so on. The Australians have also sought to predict outbreaks of mosquito-borne Murray Valley encephalitis in southern Australia using El Niño information (Nicholls, 1986).

Australians have had their share of successes and problems in forecasting El Niño episodes and, therefore, their impacts. However, as Graeme Hammer and Neville Nicholls, among others, have noted, "the value of seasonal forecasting is not a one-off occurrence . . . The potentially large benefits will accrue over time as ENSO cycles continue to occur" (Hammer, 1995, p. 5).

### *Peru and the 1997–98 El Niño*

A recent study of the use of El Niño information in Peru in four sectors of society (public safety, fisheries, agriculture and hydropower) showed that, before the 1997–98 event, of these sectors only the fishing and agricultural sectors were using El Niño forecasts to some extent in their decisionmaking processes. Interestingly, public safety organizations and hydropower agencies seem to have paid little attention to El Niño and its possible impacts on their activities (Glantz, 1998a). The fishing industry has been coping with El Niño-related impacts and research for a few decades (e.g., Flores-Palomino, 1998), and the agricultural sector seems to have developed more serious interest in El Niño in the late 1980s (Lagos and Buizer, 1992). Hydropower agencies became concerned about El Niño's impacts on its activities only after the 1991–92 event (Glantz, 1998b). To be quite honest, even President Fujimori of Peru seemed to have had little interest in the El Niño events of the 1990s, until a strong El Niño was forecast in mid-June 1997, and political considerations dictated his involvement in attempts to mitigate potentially adverse impacts.

At that time, President Fujimori created a task force made up of several government agencies with the responsibility to prepare for the devastating heavy rains and floods that were forecast for the Southern Hemisphere

Domingo 26 de octubre de 1997 *El Universo* p. 3

Según congresista peruano

**El Niño impedirá conflicto**

*Figure 7.3. "El Niño Will Impede Conflict" was the headline from an Ecuadorian newspaper,* El Universo, *26 October 1997.*

summertime (December, January, February). Fujimori was seen on Peruvian television each night at first organizing preventive works and, later, helping flood victims, assessing damage, clearing mudslides with heavy equipment, and so on. The *Economist* (27 September 1997) took note of his activities and personal involvement in preparing for El Niño's adverse impacts on Peru, in an article entitled "Fujimori against El Niño."

At the time of the 1997–98 El Niño, Peru had been engaged in a "hot" war with Ecuador since 1995 as a result of a border dispute. Newspapers in Peru and in Ecuador noted that it was likely that there would be no attack on Ecuador by Peruvian armed forces, because the heavy rains that accompany El Niño would have made the terrain impassable and the cloud cover too thick for air strikes (*El Universo* [Ecuador], 1997; Figure 7.3).

Various groups around Peru held meetings and workshops in order to educate themselves on El Niño's potential impacts and how they might prepare for them: clearing irrigation ditches, sewers, dry river beds, and so on, to enable the torrential rain water to pass swiftly to the ocean. For the first time, Peruvians in all walks of life and at all economic strata were focused on El Niño. A major Peruvian television station began to carry nightly weather updates and even created an El Niño Early Warning System (called SATEL in Spanish) (A. Levy, personal communication, 1999).

The impacts on Peru and its economy were on the order of US$3 billion dollars. It is difficult to calculate the cost of what that El Niño might have been had the President not mobilized Peruvians to prepare for it. One must conclude that, at least qualitatively, there was benefit to Peruvian society as a result of the combination of the mid-1997 NOAA forecast of a strong El Niño and Fujimori's (not just government agencies') response to it. Nevertheless, impacts were still of major proportions and very costly (Zapata-Velasco and Sueiro, 1999).

## Some examples of missed forecasts

Along with several success stories of country responses, there have been some notable misses with regard to forecasting El Niño or its impacts in the past 25 years. These examples put a realistic limit on our expectations about the value of El Niño information to our decisionmaking needs.

*Quinn–Wyrtki forecast: 1974–75*

Early in 1974, oceanographers William Quinn and Klaus Wyrtki believed that on the basis of their research about the phenomenon, an El Niño was likely to develop in early 1975. This took place in the period of research focusing on coastal upwelling and on El Niño processes in the equatorial Pacific, when less was known about them. Quinn and Wyrtki's projection was supported by other researchers in the Fall of 1974. On the strength of those beliefs, the US National Science Foundation funded, on very short notice, two research cruises to be carried out in the eastern equatorial Pacific in 1975 between February and May (Wyrtki *et al.*, 1976).

Observations taken during the first cruise reinforced the Quinn and Wyrtki projection about the likelihood of the further development of an impending El Niño. The second cruise undertaken a few months later, however, yielded observations that those emerging El Niño conditions had suddenly changed into a cold event, a phenomenon in which at that time there was little, if any, scientific interest. The field program was then canceled and the ships returned to port. Their forecast of El Niño was considered to have failed. Some researchers argued at the time that, although a full El Niño did not develop in 1975, new insights into oceanic processes were gained. For example, although El Niño-like processes may appear in the eastern Pacific they may not unfold into fully developed El Niño events (Wyrtki *et al.*, 1976).

Wyrtki summarized his view of their forecast and the "collapsed" El Niño as follows:

> The year 1975 will not enter oceanographic history as a year of a large El Niño. However, as predicted, an El Niño situation started to develop with a characteristic overflow of warm, low salinity water from the north, an intensification of the undercurrent, and an accumulation of sub-surface water along the coast [of Peru]. Without the El Niño expedition [the two cruises], these conditions would not have been observed, and from coastal temperatures alone one would have concluded that nothing abnormal had happened. This investigation indicates that only very strong El Niño events have been recorded in the past and that weak occurrences have remained unnoticed.
>
> (Wyrtki *et al.*, 1976, p. 346)

Interestingly, an El Niño event did develop a year later in 1976. It is only fair to note that modelers who try to reproduce anomalies that have occurred in the past in order to verify their models have typically had difficulties in replicating El Niño events in this period. One researcher suggested that Quinn and Wyrtki were unlucky to have started to forecast El Niño with the toughest case in the past 25 or so years (M. Cane, personal communication, 1996).

### *The 1982–83 El Niño*

As noted in Chapter 6, most El Niño researchers, with a couple of exceptions, failed to forecast the onset of this major event.

### *A 1982–83 El Niño teleconnection impact forecast*

Another notable problem with regard to a forecast of El Niño's impacts took place in early 1983. A university professor and colleague issued a forecast of the impacts on agriculture in the US Midwest of the El Niño that began in late 1982. They projected, on the basis of about 100 years of El Niño and corn yield records, that corn yields in Illinois would be extremely high. Their projection was given very high visibility after it was published in the journal *Science* (Handler and Handler, 1983). It was also carried in major American newspapers. However, the projection of corn production proved erroneous (i.e., "dead wrong") when, later in the season, it became clear that corn yields had actually been reduced by about 50%. For a while, this erroneous forecast undermined the credibility among scientists of such teleconnection forecasts.

## Forecasts in the 1991–95 period

The first half of the 1990s proved to have been somewhat of a nightmare for the El Niño forecasting community. Things seemed to go well at the very beginning of the decade. The onset of an El Niño in late 1991 was correctly forecast. However, the demise of this event, which had been expected to take place by the end of 1992, did not occur after sea surface temperatures declined from warm to near normal. A weak El Niño then began to develop in the eastern equatorial Pacific in 1993, and once again seemed to die out, only to be followed by yet another event in 1994.

When one looks at the sea surface temperature anomalies in the first half of the 1990s for the different Niño regions of the equatorial Pacific, as shown in Figure 4.4), it becomes difficult to gauge the El Niño situation, even with hindsight. This is because of conflicting signals coming from the different parts of the ocean. Figure 7.4a shows sea surface temperature anomalies in Niño1 and Niño2 regions (along the western coast of South America). One can see three distinct, albeit not very strong, El Niño-like warmings in early 1992, in early 1993, and in late 1994.

Figure 7.4b shows anomalies for the Niño4 region in the western

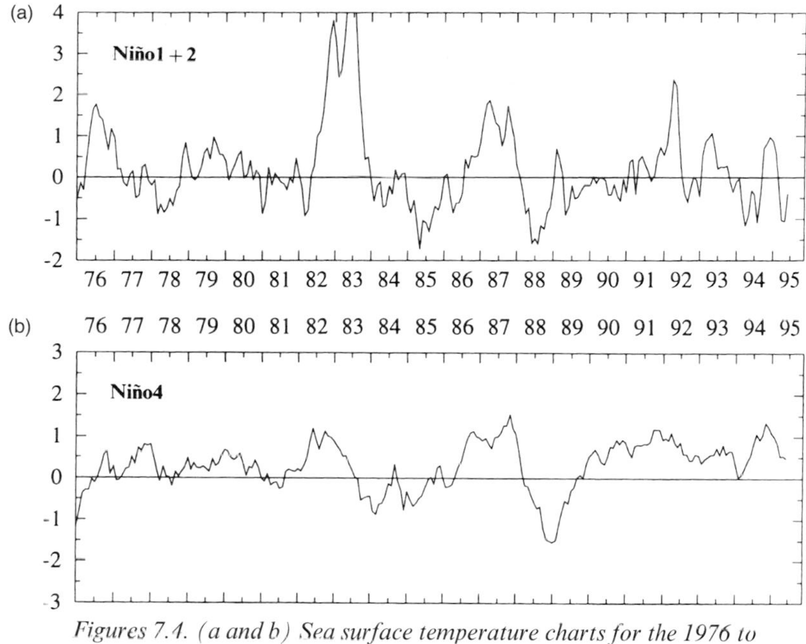


*Figures 7.4. (a and b) Sea surface temperature charts for the 1976 to 1995 period for Niño1 + 2 regions compared to the Niño4 region. The comparison suggests that an El Niño occurred in the western Pacific from 1989 to 1995, while three El Niño events of varying intensities appeared in the eastern Pacific. (From CAC, 1995.)*

equatorial Pacific for the same period of time. One can see that the sea surface temperatures stayed above average from late 1989 to early 1995. Researchers monitoring the Niño4 region could argue that one long intense El Niño occurred in the early 1990s, while those monitoring the Niño1 + 2 regions believed that three shorter and weaker El Niño events occurred then.

This El Niño confused the forecast community because, depending on the region on which one focuses, the event could be viewed either as a long event or as a set of three smaller ones. Yet another interpretation could be that an El Niño event occurred in 1991–92 in the eastern equatorial region, followed by occasional weak, above-average increases in sea surface temperatures (but not El Niño).

The US National Weather Service (NWS) created a program in early 1995 to produce long-range forecasts. Using monitoring and modeling techniqes and a variety of other methods, the NWS issues forecasts each month for periods of up to 15 months into the future. Sea surface temperature changes in the equatorial Pacific play a central role in constructing their forecasts.

The NWS picked a difficult period in which to begin its operational long-range forecasting activity – in the midst of such unusual behavior of sea surface temperatures across the equatorial Pacific Ocean. As one science writer noted at the time,

> the El Niño now developing [December 1994] came as something of a surprise . . . It is the third in four years, and El Niño forecasts, including that of the NWS's coupled model, didn't see it coming until late summer [1994]. With El Niño so central to extended-range forecasting, that's unsettling some people.
>
> (Kerr, 1994, p. 1941)

### *Australia's seasonal outlooks in the 1990s: hits and misses*

The decade of the 1990s opened with the promise for Australian scientists to consolidate their routine seasonal outlook system, which began operations in 1989. This system was based on statistical assessments and an understanding of the "phases" of El Niño. The detail of, and confidence in, the system was reinforced by a greatly enhanced international climate monitoring capability and by improvements in modeling the dynamics of sea surface temperature and Southern Oscillation behavior in the equatorial Pacific. That promise was partly fulfilled in the 1991–95 period.

Highlights of the successes and failures of the seasonal outlooks issued by Australia's Bureau of Meteorology in this period were as follows (note that the seasons referred to are those in the Southern Hemisphere, which are opposite to Northern Hemisphere seasons: summer is winter, Fall is spring, and so on):

- *July 1991*: The first successful operational forecast was issued of an El Niño-related drought in parts of Australia. The winter and spring conditions in eastern Australia were well predicted
- *Austral Autumn 1992*: Based on the usual end of Australian droughts in the Southern Hemisphere autumn, the community was advised that drought was likely to end with this event as well. This, however, was not a useful outlook for New South Wales and Queensland, where drought persisted
- *Spring 1992 and Summer 1992–93*: The SOI rose to near zero and was accompanied by unpredicted extremely wet conditions in southeast Australia
- *First half of 1993*: Dry conditions reappeared and predictions were in part successful
- *Austral spring 1993*: Another apparent breakdown in the El Niño episode. Predictions from June to the end of the year were both on target and useful to the farming community
- *1994*: There was a re-emergence of a major El Niño. Predictions from June

to the end of the year were both on target and useful to the farming community
- *Early 1995*: Cautious statements were issued on the probability that the 1994–95 El Niño episode would end. Such statements were well received by the government and the rural sectors. Good rains were received over much of eastern Australia through 1995

Australia's Bureau of Meteorology issues rainfall forecasts referred to as "Seasonal Climate Outlooks". These outlooks were issued each month throughout the 1990s. The outlooks issued in late 1997 generated a heated debate over the interpretation of the wording used in the forecasts. Tahl Kestin and Neville Nicholls (1998) assessed the forecast-response situation (see also the box on p. 200) and provided an Australian example of forecast use in bush fire prevention.

Whenever climate-related forecasts are issued, officially or unofficially, operationally or experimentally, and appear in either the scientific or popular media or on the Internet, some people are listening. Some of those people will take action based on those forecasts, and they will either reap the benefits or suffer the consequences for doing so.

A study of a bad (missed) forecast of seasonal water supply in an agricultural region in the US state of Washington raised an interesting question about the value of a long-range forecasting: "could the costs incurred because of a bad seasonal forecast balance out the hypothetical value of a larger number of good forecasts?" (Glantz, 1982). A similar study of an incorrect forecast could be done for the forecast of drought in southern Africa as a result of the 1997–98 El Niño. Instead of the regional drought that had been forecast, it rained. To assess the true value to society of an El Niño-related forecast system, one must include the costs of missed forecasts in the overall calculation of the benefits.

The examples of missed forecasts are not presented here to make light of the serious research efforts to develop a reliable long-range (including El Niño) climate forecast system. They are presented in order to underscore the reality that forecasting El Niño is wrought with difficulties and surprises. Although progress has been made in the 1990s, and although forecasts are being used increasingly, a totally reliable El Niño forecast is still not in hand. Despite the marginal success of forecasts for the 1997–98 event, there is strong pressure from government funding agencies and the modeling groups they fund to describe to the public existing El Niño forecast skills in only very positive terms.

While attempts to forecast El Niño are worthwhile research efforts, deserving of major scientific research funding by governments around the globe, the state of scientific understanding must not be oversold to the public and to policymakers as they have been in recent times (e.g., NSF, 1998). Scientists should label their forecasts as experimental and should include warnings about the potential misuses of a probability-based El Niño forecast. In fact, bulletins of the South African Weather Bureau's Research Group for Statistical Climate Studies (RGSCS) included such an appropriate disclaimer warning that the users of their information beware. Their disclaimer was as follows:

> The RGSCS and the South African Weather Bureau accept no responsibility for any application, usage or interpretation of the information contained in this document and disclaim all liability for direct, indirect, or consequential damages resulting from the usage of this bulletin.

Some forecasters have clearly had more success than others. Yet, no forecast group has an unblemished El Niño forecast record. Sir Gilbert Walker once issued words of caution about making projections related to weather phenomena. In his 1935 address on "Seasonal weather and its prediction," he chose to warn his colleagues about the dangers associated with long-range forecasting. He noted that

> some of the most progressive countries in the world are inclined to make predictions on an insecure basis; their technical staff does not realize that, though the prestige of meteorology may be raised for a few years by the issue of seasonal forecasts, the harm done to the science will inevitably outweigh the good if the prophecies are found unreliable . . . It is the occasional failures of a government department which are remembered. (Walker, 1936, p. 117)

Walker's words of the 1930s are still valid today. It was noted earlier that forecasters in FUNCEME in the Brazilian northeast had a few successful forecasts, for which the organization received high praise in national as well as international circles. However, following FUNCEME's three successful interannual forecasts in the early 1990s, a subsequent seasonal forecast by FUNCEME was perceived by the public to have been in error. As a result, the credibility of the organization, its forecasters, and its forecasts was sharply eroded. The process to restore the high level of credibility to FUNCEME began almost immediately with the help of other Brazilian research groups.

While FUNCEME ended up taking the heat, other key confounding factors had to be considered: the El Niño event(s) of the early 1990s were among the least predictable in recent times. In addition, the previous three successes of FUNCEME led the public to feel that the forecasts were going to be absolutely correct. The public did not listen carefully to the caveats (words of caution) associated with the forecast. As a result, they came to expect too much perfection from a forecast based on probabilities.

As another missed forecast example, the Cane and Zebiak models (called LDEO1 and LDEO2), which had been so successful in 1986 and in 1991, had forecast a cold event (La Niña) for 1997–98. They did so even as the waters in the central Pacific were heating up rapidly. Other El Niño modelers and forecasters have suggested that they had delayed issuing their forecasts of El Niño's possible onset because the El Niño flagship model at the time – the Cane and Zebiak model – contradicted their own models' output. For whatever reason, once credibility has been damaged, it can be a long-term process to get it back.

The international research community now believes that it is on the threshold of producing reliable operational forecasts of El Niño and La Niña on a routine basis. In the mid-1990s, the US Government's NOAA, Columbia University and the University of California at San Diego established the International Research Institute (IRI) for climate prediction. The IRI's long-range mission has been to assess and develop seasonal-to-interannual forecasts on a continuous basis. That mission also includes the fostering of the application of such forecasts to the explicit benefit of societies worldwide. The IRI works on improving model and forecast system development, experimental forecasting, climate monitoring and dissemination, research on applications, and training. These activities are supposed to be carried out in collaboration with the international climate research and applications community. The Institute will eventually assume multinational governance. Although much of its early work has focused on ENSO-related climate forecasts, its responsibilities encompass non-ENSO-related climate forecasts as well. Originally, this institute was expected to be linked to regional forecast application centers in various parts of the world. Regional centers would then tailor the forecasts on various time scales with various lead times to local and regional conditions. A 5-year review of the IRI has brought about some changes. The activity is currently centered at Columbia University, and Scripps is no longer a partner in the venture. Another major change has been in emphasis, with less effort going to the generation of a seasonal consensus forecast and more emphasis on the application of forecasts to match user needs.

### *Comparing forecast use: a three-country example*

The 1997–98 El Niño was the most monitored natural event in history. Unlike during the 1982–83 El Niño, the scientific community, governments, industries, and individuals became aware of El Niño's onset, at least within a few months. They were then made aware of the potential impacts that might accompany this event. Although governments were armed with an El Niño forecast by June 1997, each seemed to respond differently, even in locations where the forecast of El Niño's teleconnections had become reliable. The following few paragraphs provide a brief description of three countries that had received El Niño forecasts at about the same time.

> The operational prediction of climate fluctuations promises rewards as rich as those of operational weather prediction. A major difference between weather and climate is that the one involves primarily the atmosphere, while the other is a product of the atmosphere interacting with the various water, land and ice surfaces beneath it. Climate prediction requires the concerted efforts of atmospheric scientists, oceanographers, biologists and hydrographers who jointly develop coupled atmosphere–ocean–ice–land models capable of simulating and forecasting climate fluctuations. The activity, because it provides the public with invaluable information, will justify a network of instruments that measure, on a routine basis, not only the atmosphere but also the oceans and land conditions.
>
> George Philander, Princeton University, 1996

The governments of Peru, Kenya, and Costa Rica had received forecasts by June 1997 from their official national and international meteorological sources that a strong El Niño was developing in the central Pacific. In Peru, President Fujimori established an interagency task force to focus on El Niño. Its purpose was to plan pro-active measures to mitigate the potential (but highly likely) adverse impacts on agriculture, fishing, and public welfare. The government eventually sought financial assistance from the World Bank and received about US$150 million to support its efforts to "combat" El Niño (Stolz and Sanchez, 1998). Adverse climate conditions for northern Peru during El Niño is quite a reliable relationship. Costa Rican drought, too, has relatively reliable teleconnections to El Niño events.

By June 1997, Costa Rica's National Meteorological Service provided its government with an El Niño forecast. Government agencies prepared a detailed informative report to the World Bank to support their request for an El Niño-related loan. The plan was completed in August 1997 (CORECA, 1997). The World Bank, however, denied financial support to Costa Rica in advance of El Niño, offering to provide loans after there were adverse impacts on the country. In response to the Bank's loan denial, the Costa Rican government played down the fact that a major El Niño was brewing in the tropical Pacific, reasoning that the government did not have the resources to take the necessary preventive measures to mitigate El Niño's impacts and would be unable to meet demands from its citizens for funds to prepare for El Niño's impacts.

Kenya's National Meteorological Service and the Regional Drought Monitoring Center which is located in Nairobi issued an El Niño forecast in June 1997. The Government of Kenya, therefore, was sufficiently warned in a timely fashion of an emerging strong El Niño. The Kenyan government, however, did not act immediately on the forecast (as Peru and Costa

### What are the effects of ENSO in the Philippines?

In the Philippines, drought events are associated with the occurrence of ENSO episodes. The worst in spatial terms was in 1982–83. The biggest damage was caused by the 1990–92 event which was estimated to cost the country about P4.1 billion in crop and other losses.

### What are the other expected impacts of ENSO-related drought events in the Philippines?

These include a number of *environmental impacts* (degradation of soil which could lead to desert-like conditions. If persistent, affects water quality like salt water intrusion, high forest/grass/bush fire risk, domestic water supply shortage, etc.), *social impacts* (disruption of normal human activities, migration to urban communities, human and health problems, etc.), and *economic impacts* (unemployment, food shortages, etc., reduction in productivity and subsequent revenue of various industries, hydroelectric power generation, etc.).

### What are the typical regional impacts of an El Niño event?

By studying past warm and cold episodes, scientists have discovered precipitation and temperature anomaly patterns that are highly consistent from one episode to another. Within the tropics, the eastward shift of thunderstorm activity from Indonesia into the central Pacific during warm episodes usually results in abnormally dry conditions over northern Australia, Indonesia and the Philippines in both seasons.

### What challenges are posed to national and local governments of countries affected by ENSO?

- Prevent and mitigate losses which may have been caused by the drought for crops planted through 1998. Drought tolerant crops suitable to soil conditions in the Philippines include the following:

  Sorghum, Sweet pepper, Ube, Monggo sprouts, Alugbati, Banana, Winged bean, Pigeon pea, Cashew, Mango, Pomelo, Citrus, Tamarind, Avocado, Jackfruit, Guava, Pineapple, Ginger, Cassava, Cantaloupe, Asparagus, Black pepper, Sweet potato, Monggo bean, Grapefruit, Cucumber, Cowpea, Peanut, Grape

- To encourage planters and farmers to adjust their planting schedules to avoid damage from floods and typhoons from July 1998 onwards. Should prolonged dry spells occur, coordinate with disaster and relief agencies for multisectoral approaches and adoption of mitigating measures.

DOST PAGASA: (Department of Science & Technology, Philippine Atmospheric, Geophysical and Astronomical Services Administration)

Rica had done). Perhaps government officials perceived the reliability of the teleconnections as low between El Niño in the tropical Pacific and rainfall in East Africa. However, unnoticed at the time was that the western Indian Ocean was also unusually warm. This alone increased the chance for heavy rainfall along the east coast of Africa. Later in the year, as the heavy rains began to take their toll on life and property, the Kenyan government began to respond to El Niño's impacts (i.e., react instead of pro-act). At first government authorities blamed the destruction of major roads and bridges on El Niño. Other observers noted, however, that the country's infrastructure was in a state of disrepair and near collapse from decades of neglect. Nevertheless, perhaps the Kenyan government should not be faulted for its delayed response to the El Niño forecast of June 1997, because El Niño's impact on East Africa is not so clear. Kenya, like Peru and Costa Rica, also appealed to the World Bank for El Niño-related financial assistance and received it under the condition of close scrutiny by the Bank to minimize the loss of disaster funds to corruption.

Forecasts of El Niño now abound. They are issued by government agencies in the USA, Canada, Australia, Japan, South Africa, the UK, and Germany among other countries, as well as by consulting firms in various sectors of society and by different research groups as well. Today, societies are being increasingly forewarned and, as the saying goes, forewarned *should* mean forearmed. However, this is not always the case. In the light of our improved scientific understanding of the El Niño phenomenon over the past three decades, one might realistically assume that the cost (in terms of constant dollars) of the impacts of the 1986–87, 1991–95, and 1997–98 events should be decreasing, all things being equal. But all things are not equal; for one thing, El Niño intensities vary widely. Furthermore, some ENSO researchers and environmentalists now suggest that El Niño events will increase in frequency and intensity as a result of global warming of the Earth's atmosphere (e.g., Trenberth, 1999).

Until now, there has been an overemphasis by researchers on developing a capability to forecast El Niño's onset and, at the same time, an underemphasis on using much of the El Niño information that already exists. While improving its forecasts of El Niño and La Niña, the research community must help society to figure out how best to use the El Niño information that it has already produced. Society would then be better able to adapt to, if not prevent, the worst consequences of future El Niño-related climate anomalies. There is no doubt that El Niño information (and I do not mean only forecasts) has tremendous *potential* value to society. Until recently, support for research on the societal aspects of El Niño was ignored by funding agencies. That must change if society is to turn potential benefits into real ones.

# 8 Forecasting the 1997–98 El Niño

## An overview of the 1997–98 El Niño/La Niña

In the Executive Summary of the *La Niña Summit* report (Glantz, 1998b), Michael McPhaden of NOAA's Pacific Marine Environmental Laboratory (PMEL) summarized the El Niño progression in 1997–98. He noted the successes of the TOGA-TAO Array in monitoring the evolution of the 1997–98 El Niño, and those successes were in stark contrast to 15 years previously when the 1982–83 El Niño had not even been detected until it was nearly in its peak phase. The recent monitoring successes were the result of: (a) the development of an observing system of satellite and ocean measurements, anchored by a buoy network spanning the equatorial Pacific providing data in real time (within hours of collection); and (b) the development of computer forecast models for predictions.

McPhaden also noted that the onset of El Niño was heralded by warm sea surface temperature (SST) anomalies (i.e., deviations from normal) erupting in the tropical eastern Pacific during April–June 1997. By July 1997, SST anomalies in this region were the warmest observed in the past hundred years. By the end of the year, SST anomalies exceeded 5 degrees Celsius, and were higher than those observed during the 1982–83 event (up to that time, the strongest of the century).

As the warm SST anomaly began to wane in early 1998, prediction models were largely consistent among themselves in forecasting an end to El Niño, and in forecasting the development of a La Niña in the second half of 1998. However, El Niño researchers were surprised at how quickly the tropical Pacific switched from warm to cold conditions. McPhaden commented (Glantz, 1998b) that a temperature drop of 8 degrees Celsius at one buoy location in the central Pacific over a relatively brief 30-day span had not been observed before, making the decay of this El Niño as surprising as its relatively rapid onset. He noted that the "seeds" for El Niño's decay, and potentially for the onset of a La Niña, were to be found below the ocean's surface. In late 1997, the buoy

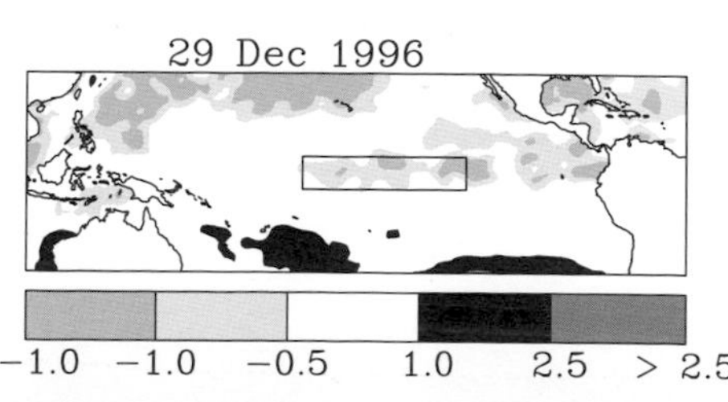


*See p. x for explanation of graphics.*

array detected a cold mass of water – located at a depth of 100–150 m west of the Date Line – and documented its expansion eastward along the equator.

McPhaden (personal communication) observed that, during the 1997–2000 period, there had been a spectacular display of climatic variability. In the first part of that display, El Niño ended and in the second part, La Niña began. The completed TOGA observing system, designed specifically for ENSO detection and forecasting, provided essential information at a level of detail never before possible. As a result, the climate "crystal ball" of the monitors was sufficiently clear in 1997 to motivate national and local governments to undertake disaster preparedness (mitigative) efforts and other responses worldwide to El Niño conditions that were developing on an unprecedented scale.

McPhaden underscored the following key aspects of the 1997–98 event:

- The successes in observing the evolution of the 1997–98 El Niño, and forecasting some of its global impacts, were in striking contrast to the situation only 15 years earlier when the major 1982–83 El Niño had not even been detected until it was nearly at its peak.
- As the warm SST anomalies began to wane in early 1998, prediction models were largely consistent among themselves in forecasting an end to El Niño, and the development of a La Niña sometime during the second half of 1998.
- The seeds for El Niño's demise were found below the ocean's surface.
- The climate "crystal ball" was sufficiently clear in 1997 to motivate many governments to undertake, on an unprecedented scale, timely disaster mitigation efforts.

(Glantz, 1998b, p. 3)

### *An overview of the El Niño forecasts for 1997–98*

I do not often write in the first person, but this particular topic calls for it. Unaware of each other's activity in the spring of 1998, Tony Barnston, a statistically oriented climate forecaster for NOAA's Climate Prediction Center (CPC) in Camp Springs (Maryland) and I had decided independently (and unbeknownst to the other) to review the forecasts of the 1997–98 El Niño event. In the midst of our separate reviews, we each learned that the other was undertaking a similar assessment. Barnston was making a quantitative review of the forecasts and I, a social scientist, was undertaking a qualitative review. In mid-1998, we decided to work together to assess the numerous forecasts of the onset of the 1997–98 El Niño event. The truth of the matter is that, although we were working on the same issue – how good were the forecasts of the onset of the 1997–98 El Niño – each of us had decided to undertake a review for different reasons.

The purpose of Barnston's review was to evaluate the forecasts produced by various computer models in order to identify the current state of the El Niño experimental forecasting models (i.e., their level of reliability). More specifically, Barnston wanted to see whether one type of model was performing better, the same, or worse than the other type (i.e., dynamical models (representing the physics of the ocean and the atmosphere) as opposed to statistical models (using historical data only)). He also wanted to assess the level of skill of these model types. He had undertaken a similar study a few years earlier looking at experimental and operational El Niño forecasts during the 1982–93 period (Barnston *et al.*, 1994). His new review was going to identify any improvements since then, given the completion, of the TOGA-TAO monitoring system, and the recent advances in modeling the ENSO cycle of warm and cold events.

My own reasons for reviewing the 1997–98 El Niño forecasts were much less scientific. I wanted to find out whether the growing number of positive statements about the success of the 1997–98 El Niño and El Niño-related forecasts made by government agencies, modelers, and the media were as valid as scientists had been claiming. My recollection was that the forecasts of the onset of the 1997–98 event did not closely match observations. I hoped that my review would help to sort out who had said what, when and how accurately about that El Niño's onset. Many of these statements of success suggested that the researchers had forecast El Niño several months in advance, thanks to the "big" computer models (Kerr, 1998; NSF, 1998). In other words, I wanted to see how close to reality the comments had been of the forecasters and modelers who were being widely quoted by the print and electronic media.

For my part of the joint review, I focused on the highly visible contents of the executive summaries of a bulletin then published quarterly by NOAA's CPC. This bulletin – the *Experimental Long-Lead Forecast Bulletin* (ELLFB) – contained a set of experimental forecasts produced by various modeling groups from around the globe and was distributed to about 1500 subscribers. It is important to note that Barnston had been the editor of the *Bulletin* until December 1997. Since 1998, the *Bulletin* has been published by the Center for Ocean, Land and Atmosphere (COLA ) in Calverton, Maryland. The ELLFB has also been published on the Internet for all to see.

The time period encompassed by the forecasts under review preceded and followed the *onset* of the 1997–98 event (March 1996 to December 1997). It is important to note that not all modeling groups working on El Niño submitted their model's projections for publication in the ELLFB. Of the 19 experimental forecasts noted in the *Bulletin*, 15 were chosen for our joint assessment (see Table 8.1). This assess-

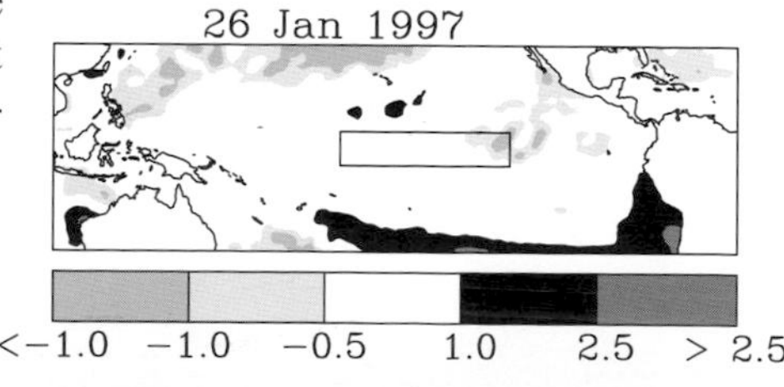

Table 8.1. *Models included in our evaluation of the 1997–98 El Niño forecasts*

*Dynamical models*
1. Lamont–Doherty Earth Observatory (LDEO) simple coupled model
2. Second version of Lamont simple coupled model (LDEO2)
3. Australia Bureau of Meteorology Research Centre low order coupled model
4. Original version of University of Oxford intermediate coupled ocean–atmosphere model
5. Second version of University of Oxford intermediate coupled ocean–atmosphere model
6. Scripps/Max Planck Institute hybrid coupled model
7. Center for Ocean–Land–Atmosphere (COLA) comprehensive coupled model
8. National Centers for Environmental Prediction (NCEP) comprehensive coupled model

*Statistical models*
9. Colorado State University/AOML CLIPER regression model
10. Climate Prediction Center of NCEP constructed analog model
11. Climate Prediction Center of NCEP canonical correlation analysis model
12. Climate Diagnostics Center/University of Colorado linear inverse modeling
13. UCLA singular spectrum analysis/maximum entropy method
14. University of British Columbia''s neurological network model
15. Climate Prediction Center of NCEP consolidated forecast model

ment was reviewed by colleagues and then published in early 1999 in a scientific journal (Barnston *et al.*, 1999). The results of our assessment apparently passed the "acid test" of critical peer review.

By comparing the forecasts over time with observations of changing conditions in the tropical Pacific, we could see how the quarterly projections produced by each of the models had fared. We could then identify the "closeness" to reality of the various forecasts at different points in time and compare them with the other forecasts that had been provided to the CPC for publication in the *Bulletin*.

I analyzed the *Bulletin's* executive summaries, based on my belief that societal users of these forecasts would probably focus on these summaries to gain a glimpse of the range of El Niño forecasts at a given point in time. Users in search of more detail could refer to the full text and graphics of the forecasts. Barnston relied on the full forecast reports that had been submitted to the bulletin. As editor of the *Bulletin*, Barnston, who had prepared the executive summaries from the full reports, noted that over the years no one had questioned the correctness of his summaries or the words he used in them. Thus the overriding purpose of the joint review from my perspective was to provide a "reality check" to public statements about how well the models had performed.

The following statements summarize the findings of our quantitative aspects of the review (Barnston *et al.*, 1999).

1. Most of the models forecast some degree of warming (but not an El Niño) above normal of SSTs in the equatorial Pacific in December 1996, when the SSTs were somewhat below normal.
2. No model predicted the strength of the 1997–98 El Niño until it was already in the process of becoming very strong in the Northern Hemisphere in the late spring of 1997.
3. Neither type of model (dynamical or statistical) produced a better forecast than the other in the period under review.
4. Some models had forecast SST changes of about 1 degree Celsius as opposed to the 2.5–3 degree Celsius increase that actually occurred.
5. As of March 1997, no model had come close to predicting the size of the 1997–98 El Niño.
6. Although the models in June 1997 had forecast a big event by the end of the year, they did not forecast the extraordinary event that ultimately did occur, even though it was one of the two biggest in the twentieth century.
7. Once an El Niño did develop, a larger number of the fifteen models were able to forecast in mid-1997 its peak phase for late 1997.
8. Forecast skill for the 1997–98 event was no better than it had been for the 1982–93 period previously reviewed by Barnston *et al.* (1994).
9. The fact that most models predicted some degree of warming in early 1997 demonstrated that forecasters' skills were better than simply basing forecasts of El Niño on climate history.
10. Forecasting the tendency for certain impacts (teleconnections) in North America to take place 5 months in advance (i.e., in winter), is not difficult, relatively speaking, because the extremes of ENSO (e.g., a strong El Niño or a strong La Niña) usually develop in the Northern Hemisphere spring or early summer and last through the following winter, as was the case in 1997–98. The reliability of such forecasts, however, requires good observations of ENSO in the Northern Hemisphere summer and reliable teleconnections to various parts of the Northern Hemisphere.
11. Once the 1997–98 El Niño set in, several of the forecasts of teleconnected impacts around the globe were correct (with some notable exceptions, e.g., severe droughts did not materialize in southern Africa and the Indian monsoon season was favorable).
12. The success of the forecasts of environmental and societal impacts of the 1997–98 El Niño was noted by the media to an unprecedented extent (see Chapter 13).
13. Forecasting the environmental and societal impacts expected to occur in a Southern Hemisphere winter that occur at the same time as the onset of an ENSO extreme warm or cold event requires one to forecast across what is referred to as the "Spring Barrier". For some countries (e.g. Chile, Australia, Peru), the official declaration of El Niño's onset in June 1997 left little time for some economic

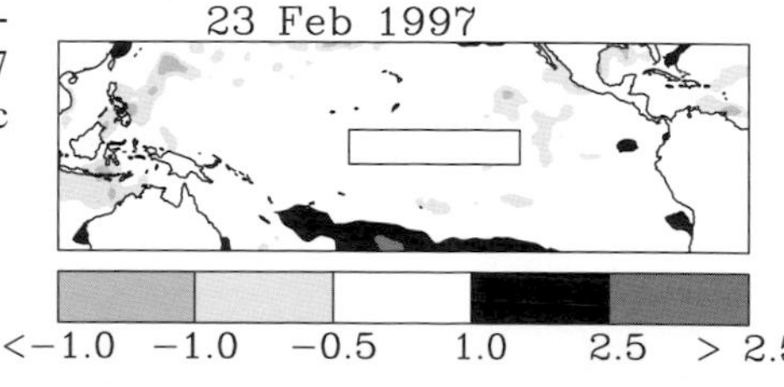

> sectors to prepare for possible adverse impacts. For example, once the onset of a major event had been confirmed in June 1997, the Peruvian government had several months to prepare for the expected flooding in northern Peru. However, its textile industry had already produced winter clothes that it could not sell because of the El Niño-related warm Southern Hemisphere winter (the Northern Hemisphere summer) in 1997.

While progress continues to be made in observing, modeling, and forecasting various aspects of El Niño, our review showed that advancements in forecasting El Niño's onset have not been as good as science reporter Kerr (1998) and news releases (NSF, 1998) had noted. Questionable statements of success such as, "The Big Models finally got it right" and "El Niño and climate easier to predict than thought," misled the public and policymakers about the state of the science and has most likely heightened expectations for improved forecasts of the next El Niño. The sought-after high level of forecasting success has yet to be achieved. Statements that praise forecast success also tend to set up the public, media, decision-makers, and even El Niño researchers, to be surprised by the next event, having been convinced by previous news headlines that the ability of the scientific community to forecast El Niño's onset has greatly improved.

## Qualitative aspects of the 1997–98 El Niño forecast

The qualitative statements contained within the individual forecasts in the *Bulletin's* executive summaries contained considerable ambiguity. For example, forecasters did not use the same terms to describe the same degree of SST change. They did not record in the same way how changes in the tropical Pacific were described. They did not use the same terminology. In addition, the various forecasts were not necessarily based on SST changes in the same tropical Pacific locations. This should not be surprising, because the forecasts came from different, sometimes competing, groups that rely on their own analysis of the tropical Pacific changes and ways of describing those changes. The chart given in Table 8.2 summarizes the forecasts produced by two dynamical models of SST changes issued from March 1996 to December 1997.

The forecasts in the chart use phrases that are not very precise and therefore not so easy for non-meteorologists to understand. For example, the word "somewhat" may provide useful information to forecasters but can be misleading to users. It suggests a change in conditions that is not very important, although when combined with temperature might suggest a trend in the change. The same can be said of the term "some warming". Is this a significant phrase and does it carry the same meaning to users as it does to forecasters?

The March 1997 Scripps/Max Planck Institute forecast uses the phrase "predicts warming through winter 1997–98". However, it does not provide any hint of the intensity of that warming. Recall that not every increase in SSTs in the tropical Pacific signifies that an El Niño will result. An El Niño is supposed to be an extreme warming, not just any warming above normal.

The June Scripps/MPI forecasts for 1997 predict "marked warmth through winter 1997–98, peaking slightly before winter". Was this a correct forecast? What is "marked"? How does one quantify it? The El Niño did occur, but it did not peak before winter. How would *you*, as a potential user of the forecast, rate this particular forecast – as a success or as a failure? It is very clear that this model did correctly foresee the development of a La Niña event for the Fall of 1998 in its September 1997 forecast.

As for the December 1996 NCEP forecast, one can ask whether the statement "becoming somewhat warm by July 1997" gets credit as a good forecast, when the SSTs had significantly warmed several months earlier. The subsequent NCEP forecasts were on the mark. However, does the phrase "strong El Niño" used in June 1997 capture the extraordinary intensity that did eventually develop?

Many of the phrases used in the forecasts of the 15 forecast systems under review were (perhaps not surprisingly) not precise. For example, what does early, mid, or late summer mean? What is meant by "through a season"? What does late summer or early spring mean? Or Fall/winter? The public has been taught that the Northern Hemisphere winter starts on December 21 and runs until March 21. To meteorologists, however, who have to deal with data compiled on monthly time frames, the Northern Hemisphere winter is viewed as DJF (December, January, February, or December 1 to February 28/9). Aside from lacking clarity with regard to the forecasts, it is very difficult to compare the accuracy or the degree of consensus about future conditions as expressed in the various El Niño forecasts.

With regard to the geographic origin of the data on which the forecasts are based, there is no apparent agreement among forecasters. For example, the various forecasts make reference to changes in the east-central Pacific in general, while others refer to Niño3.4, Niño4, or Niño3.

Some forecasters refer to changes in sea surface temperature anomalies in one of the five Niño regions, while others focus on changes in sea level pressure (i.e., changes in the Southern Oscillation Index, SOI) to discuss the development of an El Niño.

The adjectives (or qualifiers) used in the forecast statements are also quite varied and not very precise. The following examples were taken from the 15 forecasts under review: somewhat decreas-

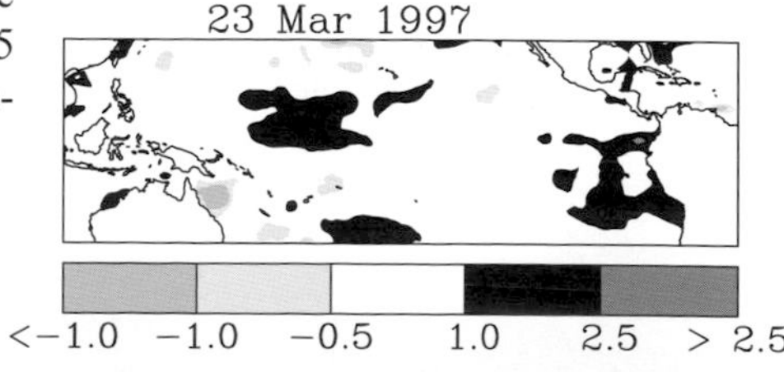

Table 8.2. *Climate change forecasts fom two dynamical models, March 1996–December 1997*

| Time | NCEP (Coupled Model)[a] | Scripps/MPI[b] |
|---|---|---|
| March '96 | Calls for some weakening of the current cold SSTA east-central Pacific between March '96 and the Northern Hemisphere (boreal) Fall '96[c] | |
| July '96 | Calls for continuing somewhat cool SSTA in east-central Pacific from now to boreal winter '96–'97, but with some warming near dateline | Predicts warm conditions for winter '96–'97 |
| Sept. '96 | Calls for continuing, somewhat cool SSTA in east-central Pacific from now to boreal winter '96–'97, but with some warming near the dateline by then | Predicts warm conditions for winter '96–'97 |
| Dec. '96 | Calls for warming through the neutral range in winter '96–'97, becoming somewhat warm by July '97 | Model predicts mildly cool conditions for winter '96–'97, moderate warming for winter '97–'98 |
| March '97 | Considerable warming through Fall '97 | Predicts warming through winter '97–'98, into spring '98 |
| June '97 | Calls for strong El Niño conditions through winter '97–'98 | Predicts marked warmth through winter '97–'98, peaking slightly before winter |
| Sept. '97 | Calls for strong El Niño conditions through winter '97–'98, moderating by summer '98 | Predicts strong warmth through winter '97–'98, dropping to normal by May '98, cold conditions by Oct '98 |
| Dec. '97 | Calls for strong El Niño conditions through spring '98, moderating by summer, becoming still weaker by Fall | Predicts strong warmth for winter to early spring '98, dropping to normal by spring–summer '98, very cold by Fall '98 |

[a] NCEP, National Centers for Environmental Prediction.
[b] MPI, Max Planck Institute.
[c] SSTA, sea surface temperature anomaly.

ing SOI; positive anomalies emerging; near-normal SOI; La Niña period peaking; below-normal SSTs dissipating; switching to warm; SOI decreases from high–normal to normal; slowly increasing to normal; coolish to normal; slightly below normal; warming slightly; warmish; slow return to near normal; slightly cool switching to slightly warm; slightly below normal SSTs neutralizing and becoming warm; somewhat cool; cool but moderating SSTs; slowly weakening to normal; warming to neutral; largely dissipating; moderating by summer; somewhat warm; neutral to slight warming; and so forth. Again, while such a wide range of imprecise statements about the state of the environment in the tropical Pacific might be understood by climatologists and oceanographers, among other specialists dealing with El Niño or La Niña, most users would have difficulty in correctly interpreting and applying such forecast statements to their specific decisionmaking needs.

## Concluding comments

Clearly, there has been an increase in the scientific understanding of both El Niño and La Niña since 1990. There have also been improvements in the way that air–sea interactions have been represented in the highly sophisticated coupled dynamical computer models. However, some scientists are overly confident about their newly discovered insights into the ENSO process. As a result of their overconfidence, they have been surprised by the occurrence of some unexpected aspect of the next event (e.g., timing, intensity, impacts). Thus the El Niño forecast success rate has not progressed as rapidly as the media and the scientists have sought to suggest.

When evaluating a model or method for its success in forecasting an El Niño or a La Niña or its societal impacts, some notes of caution are needed. For example, it is unfair to judge the value (i.e., success) of an El Niño (or La Niña) forecast based on the forecast for a single event, because a particular single El Niño forecast is unrepresentative of a model's performance or capabilities over the long run. Modelers and forecasters learn from each event and adjust their models accordingly. For example, the forecast model that had produced a few successful forecasts of El Niño since 1986 – the Cane and Zebiak LDEO2 model – had projected a cold La Niña event for 1997–98. When this model was proven by observations of Pacific SSTs to be in error, the modelers made adjustments in the way that winds in the central Pacific were represented in this model. As a result of those changes, the modelers were able to correctly forecast the 1997–98 event, *but only after the fact* (i.e., it was a "hindcast"). Another model – Scripps – that had correctly forecast a La Niña in 1998–99 had

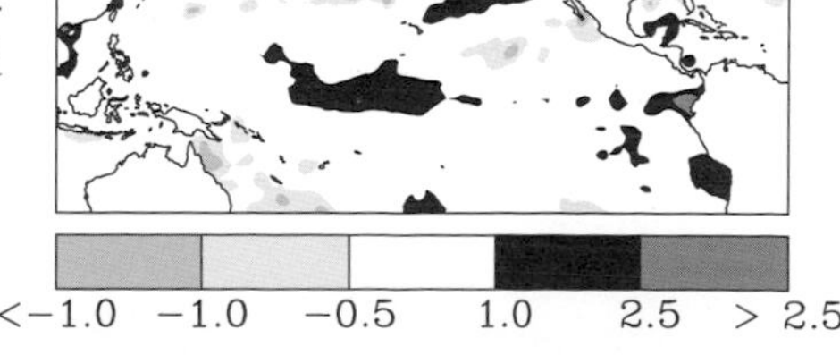

erroneously forecast an extraordinarily intense El Niño for 1996–97. Thus a model that performed poorly in forecasting the 1997–98 event may perform differently in forecasting future El Niño events, even if the model remains unchanged.

Another important point to keep in mind is that forecasters (and the media) often refer to success in forecasting El Niño – the event. However, what they are really referring to is success in having forecast some of the impacts. The science of forecasting the physical characteristics of El Niño, however, is very different from the science of forecasting the impacts (teleconnections) of an El Niño. Forecasting the 1997–98 El Niño (the phenomenon) has frequently been confused with forecasting its environmental and societal impacts. This is an important distinction. In fact, the forecasting of the 1997–98 El Niño phenomenon was not done in advance of the observations of sharp increases in SSTs in the Pacific, whereas forecasting some of the impacts was successfully done 6 months in advance.

Aside from the science, societal awareness of El Niño has sharply increased according to public opinion polls related to awareness of El Niño (CBS News, 1997). Therefore one would expect that the public and policymakers would pay increased attention when El Niño (or La Niña) forecasts are issued, especially in the next few years.

# 9 Teleconnections

Teleconnections are among the most intriguing aspects of the El Niño phenomenon. Teleconnections can be defined as linkages between climate anomalies at some distance from each other. The large distances in space and the differences in the timing between these anomalous events make it difficult for one to believe that one event (El Niño or La Niña) could possibly have influence on the other (e.g., drought in southern Africa or hurricanes in the tropical Atlantic). Nevertheless, physical and statistical research has shown that such linkages do exist. However, the influence of a climate anomaly or an El Niño on climatic conditions in a distant location depends on a variety of potentially intervening factors: how long the anomaly lasts, how intense it is, the season in which it occurs, the distance between the climate perturbations, the location of its impacts, and so forth.

A subfield of research has emerged in the atmospheric sciences that focuses primarily on improving the science behind teleconnections. Although possible linkages between two distant climate anomalies were being investigated at least as early as the mid-1800s (not merely those related to El Niño), the word "teleconnections" was apparently first used in 1935 by Swedish meteorologist Anders Ångström in his article on climate in the North Atlantic region (Ångström, 1935).

**con·nec·tion** \ kə-'nek-shən \ n [L connexion-, connexio, Fr. conec-tere]
**1:** the act of connecting : the state of being connected: as
**a:** causal or logical relation or sequence
**tele-** or **tel-** *comb form* [NL, Fr. GK tēle, tēl-, tēle, far off]
**1:** distant: at a distance : over a distance <*telegram*>
*Merriam-Webster's Collegiate Dictionary*, 10th edn, 1993

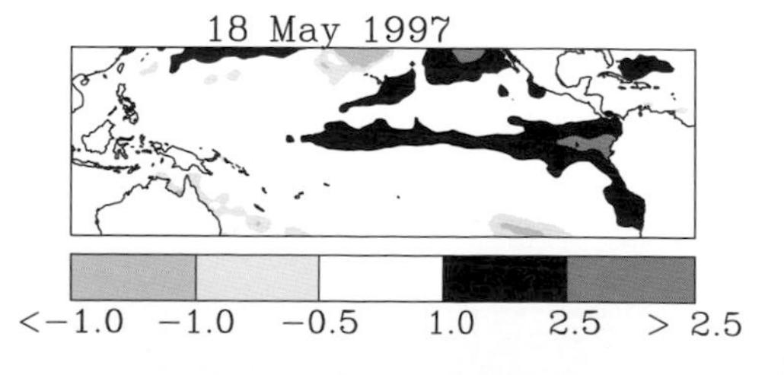

German scientists Hermann Flohn and Heribert Fleer published a thought-provoking article in 1975 on climatic teleconnections related to changes in the equatorial part of the Pacific Ocean. They sought to identify actual teleconnections in adjacent regions that, when strung together like a string of pearls, encircle the globe.

Flohn and Fleer (1975) distinguished between anomalous oceanic and climatic episodes in the central Pacific and those along the coasts of Ecuador and Peru, a distinction that became more important and more pronounced in the scientific literature in the 1980s. According to their preliminary survey, based on information from 1890 to 1973, there were a few occasions when anomalous oceanic and climatic behavior in the central Pacific was not accompanied by similar changes along the western coast of South America (e.g., 1905, 1913, 1923). Perhaps in these years the geographic distribution (i.e., location) of changes in sea surface temperatures were somewhat like the situation that appeared in the first half of the 1990s (see Figure 7.4a,b) when a sustained warming in the central equatorial Pacific was not accompanied by a sustained warming in the eastern part.

Flohn and Fleer then linked (teleconnected) El Niño in the eastern Pacific with severe drought in northeast Brazil, which was in turn linked to a decline in the level of Lake Chad, situated on the northern edge of the Sahara Desert. Streamflow runoff of the Nile River at Aswan (Egypt) was linked to changes in the level of Lake Chad in central Africa and far to the east with droughts in India. As noted earlier, there have been several attempts since the 1800s to use statistical analysis to correlate droughts in Australia with the failure of the monsoons on the Indian subcontinent. In turn, droughts in northeastern Australia have been linked directly to seesaw-like changes in the Southern Oscillation. The linkages between the Southern Oscillation and changes in sea surface temperatures in the central Pacific Ocean provided the last links in this around-the-globe chain of anomalies made up of adjacent regional teleconnections.

El Niño teleconnections also refer to temperatures. It appears that a few months after the peak in the warming of sea surface temperatures during El Niño, air temperatures increase over most of the tropics by a degree Celsius or so. Likewise, La Niña events are accompanied by cooler tropical temperatures. These temperature swings are large enough to significantly affect the global average temperature. For example, researchers noted that the 1997–98 El Niño had been responsible for part of the warming of the global atmosphere in 1998, making that year the hottest one on record.

Today, the notion of teleconnections is best known with regard to changes in the sea surface temperature in the tropical Pacific Ocean. The occurrence of a very strong El Niño brings about a shift in the locations and strengths of the high- and low-pressure systems in the tropics and extra-tropics. La Niña also generates responses in atmospheric circulation that many consider to be the opposite of those associated with El Niño. Some teleconnections related to ENSO extremes can lag in time and in space and can provide forecasters with enough lead time to identify where regional climate anomalies might occur. Lead time also provides several (but not all) societies with some opportunity to reducc El Niño's effects.

Interest in El Niño during the past two decades stemmed mainly from a scientific curiosity to understand the physical processes that underlie the Earth's climate, as mentioned earlier. However, a separate but parallel societal interest in El Niño stemmed at first from a desirc to protect guano birds, then to protect the Peruvian anchoveta fishery and fishmeal industry, and most recently to forecast climate anomalies worldwide. Today, "thanks" to the intense 1997–98 El Niño and the media's coverage of it, interest in the phenomenon and its impacts has spread to most countries around the globe. Proof can be found in worldwide national media coverage, in El Niño-related workshops, at scores of Internet websites, and in the request for proposals (called RFPs) from research-funding agencies for El Niño-related physical, biological, and social research and its application to societal needs.

In late 1997 and in 1998, climate outlook forums were held in several regions that were to be affected by this El Niño. At these forums, El Niño researchers and forecasters from the IRI met with regional and national researchers and forecasters along with potential users of El Niño forecasts to assign probabilities to various subnational regions for rainfall or temperature anomalies. Those probabilities were based on a variety of statistical and general circulation modeling activities. The maps in Figure 9.1 for Southern Africa, South America, and Southeast Asia are examples of the results of the "consensus" forecasts produced at such forums. These maps provide more detail to users of El Niño forecasts than the generalized maps of the late 1980s (e.g., Figure 9.2).

An objective of the IRI is to prepare consensus (composite) climate forecasts of a global nature based on the input of several modeling groups for distribution to regional forecast centers. Local-to-regional elements can then be factored into the forecast equation. This would tailor a global consensus forecast to regional and local conditions and needs.

Regional consensus forecasts are then constructed, evaluated and issued after a few days of deliberation at a location within the region for which the forecast was to be tailored (i.e., downscaled). Possible users of those forecasts such as agricultural planners, water resource managers, public safety and health officials, rangeland managers, and so forth are also invited to participate in the regional forums. These consensus forecasts have been experimental, but forecasters and forecast

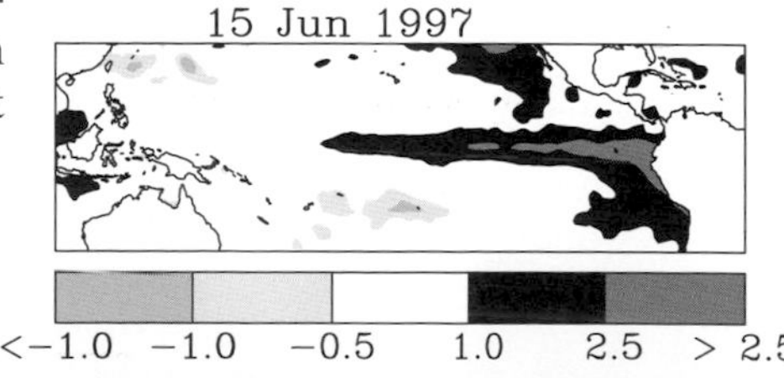

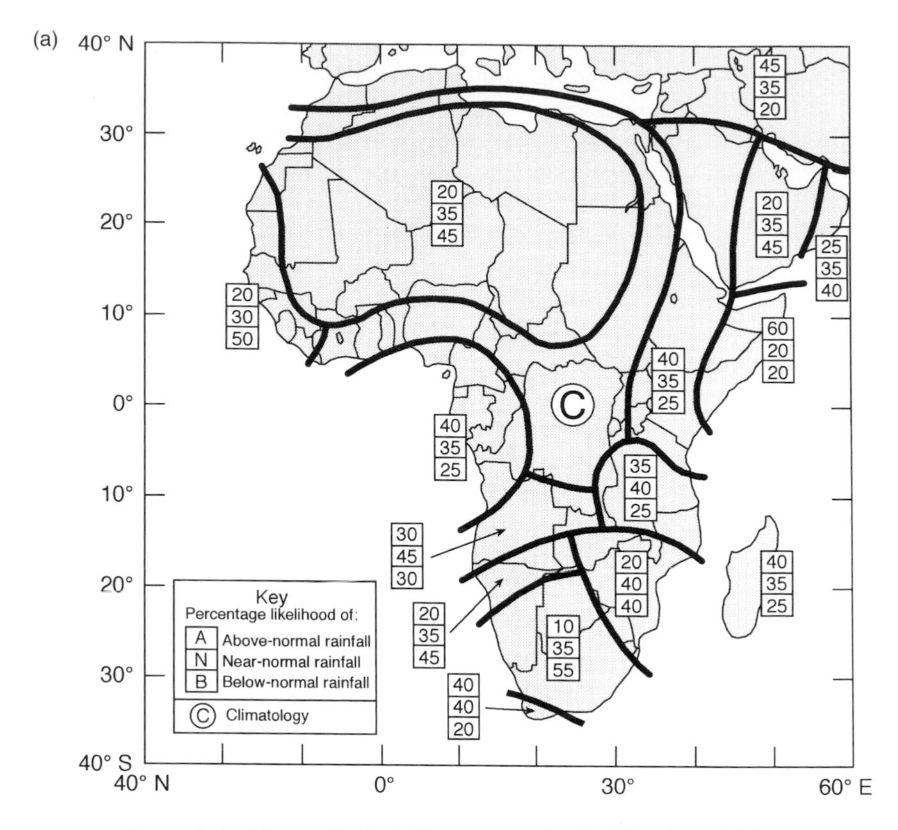


*Figure 9.1. Climate Outlook Forum maps for (a) Southern Africa, (b) South America, and (c) Southeast Asia. The letters A, N, and B in the key correspond to the top, middle, and bottom numbers in the three-number blocks shown, which are average percentages. The areas with a C are long-term averages. (Reprinted with permission of International Research Institute for climate prediction (IRI).)*

users in the various regions are interested in using these forums as a way to calibrate the reliability (and identify the uncertainties) of a seasonal-to-interannual forecast made at the global scale.

The problem of attribution (i.e., deciding which anomalies can be attributed to El Niño or La Niña with a high degree of confidence) provides scientists with a major challenge. Some anomalies give the appearance of being linked to El Niño but occur by chance with increased sea surface temperatures in the equatorial Pacific. Some teleconnections are considered to be more reliable than others and can therefore be used by decision-makers in a precautionary way.

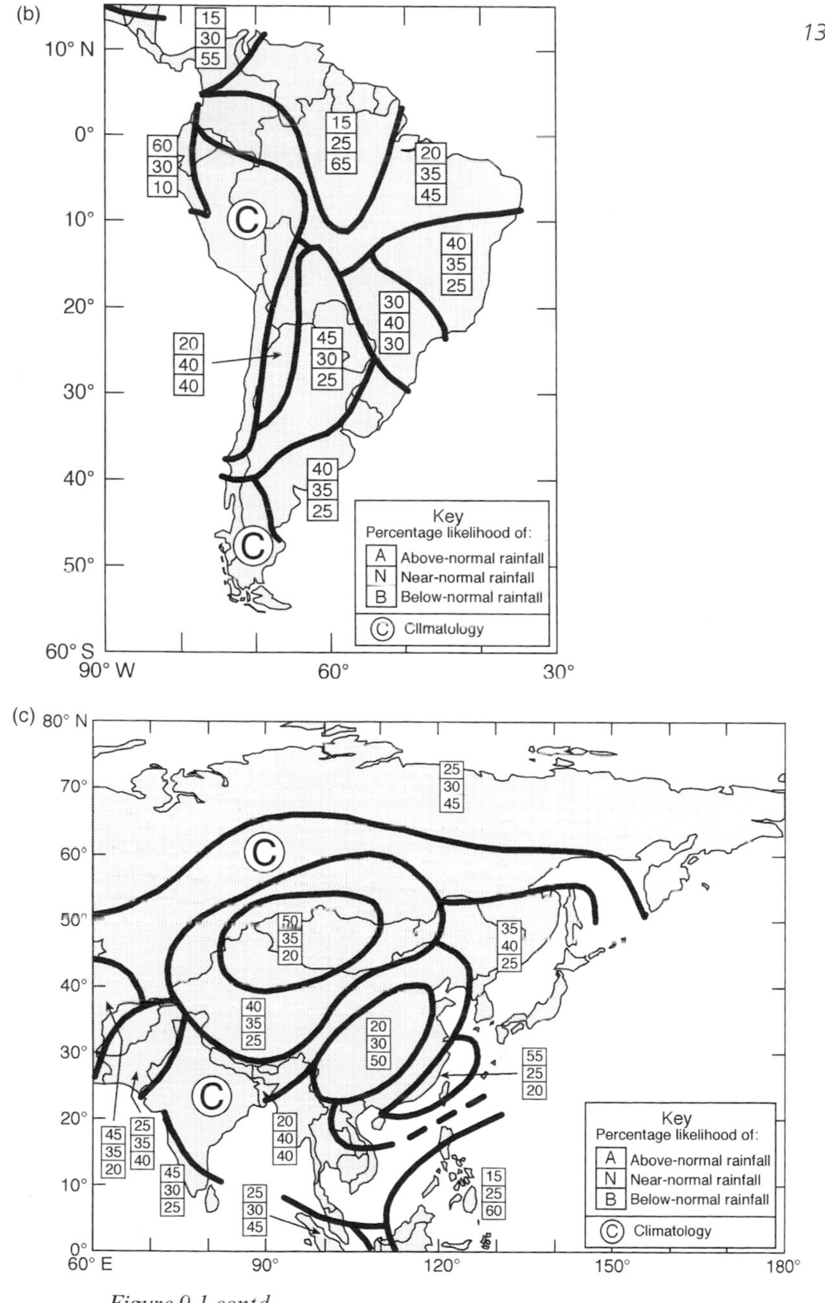


*Figure 9.1 contd.*

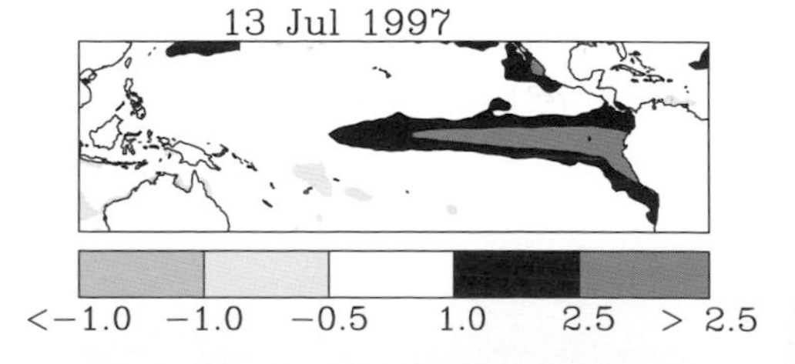

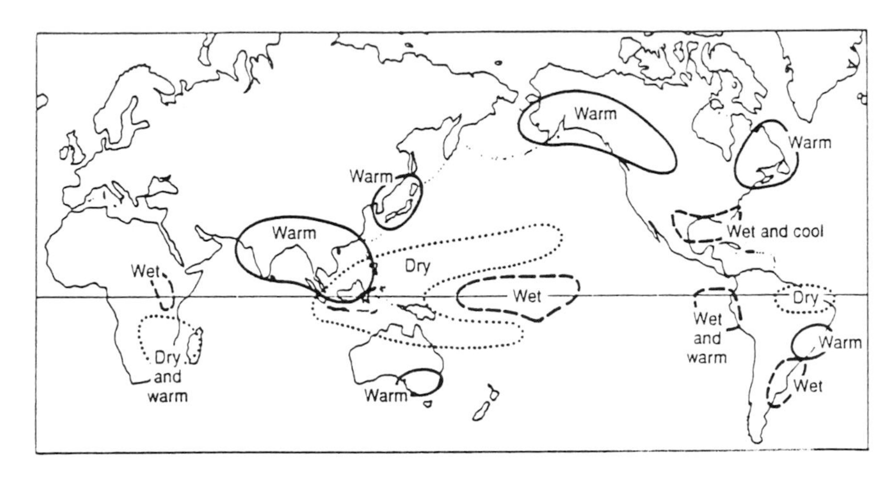


*Figure 9.2. Typical rainfall and temperature patterns associated with El Niño–Southern Oscillation conditions for the Northern Hemisphere winter season. (From Ropelewski, 1992. Website: http://www.nature.com/) Copyright 1992 Macmillan Magazines Ltd.*

Recall that, in the east–west Walker Circulation, the rising motion of the air mass has been caused by the increases in sea surface temperatures. Where the water has warmed, evaporation occurs along with the rising motion of the air and, as a result, there is an increased likelihood of rainfall in the region. The warmed air mass is carried to high altitudes toward the eastern part of the Pacific basin, where it begins to descend. Whereas the rising motion of the air generates the formation of rain-bearing clouds, its descending motion suppresses cloud formation. The Walker Circulation is from east to west at the Earth's surface and from west to east at high altitudes.

There is a similar atmospheric circulation system, called the Hadley Circulation, in the equator-to-pole direction: air rises near the equator and descends at latitudes toward the poles. The Hadley Circulation provides a connection between changes in the tropical atmosphere and the atmospheric changes outside the tropics, as shown in Figure 9.3. During an El Niño, when thunderstorm activities in the Pacific shift eastward along the equator following the eastward shift of warm sea surface temperature into the central Pacific, the jet stream (a belt of strong eastward-flowing winds) and its associated storm track also shift eastward, giving rise to El Niño teleconnection patterns in North America.

North American researchers and forecasters have become increasingly concerned with the impacts of El Niño events on North American climate, agricultural production, water resources, commerce, and public safety. For

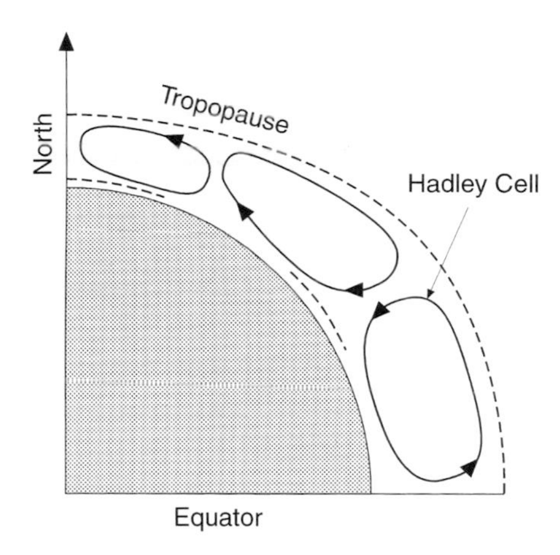


*Figure 9.3. Schematic cross-section of the Earth, showing equator-to-pole circulation in the Northern Hemisphere. (After Petterssen, 1969. Copyright 1969 McGraw-Hill, reproduced with permission of McGraw-Hill.)*

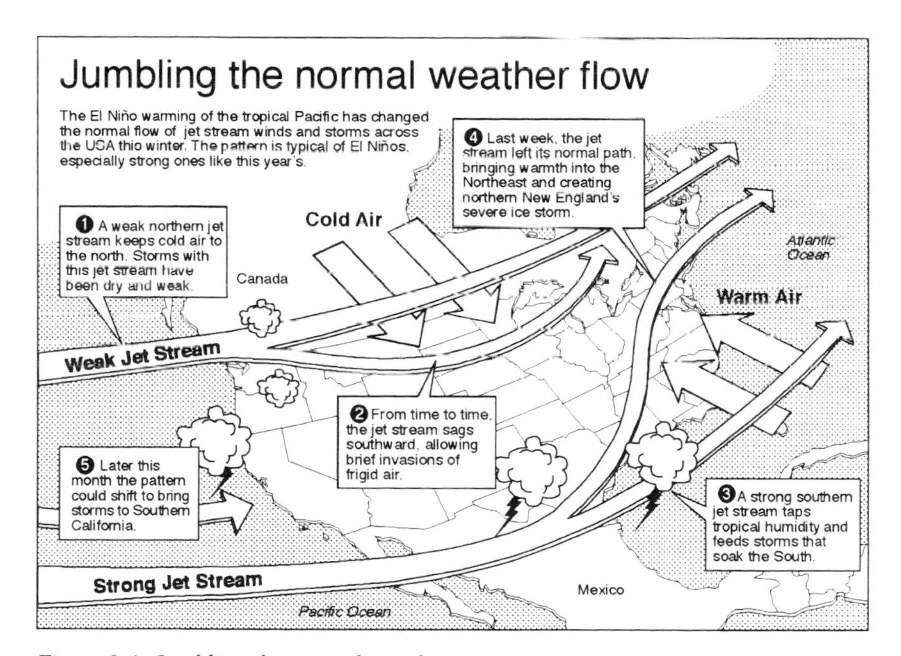


*Figure 9.4. Jumbling the normal weather flow. The effect on El Niño on the location and intensity of the jet stream over North America. (Copyright 1998, USA Today. Reprinted with permission.)*

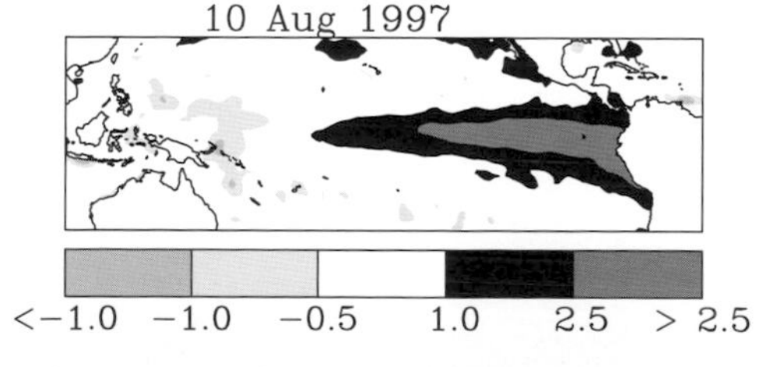

(a)

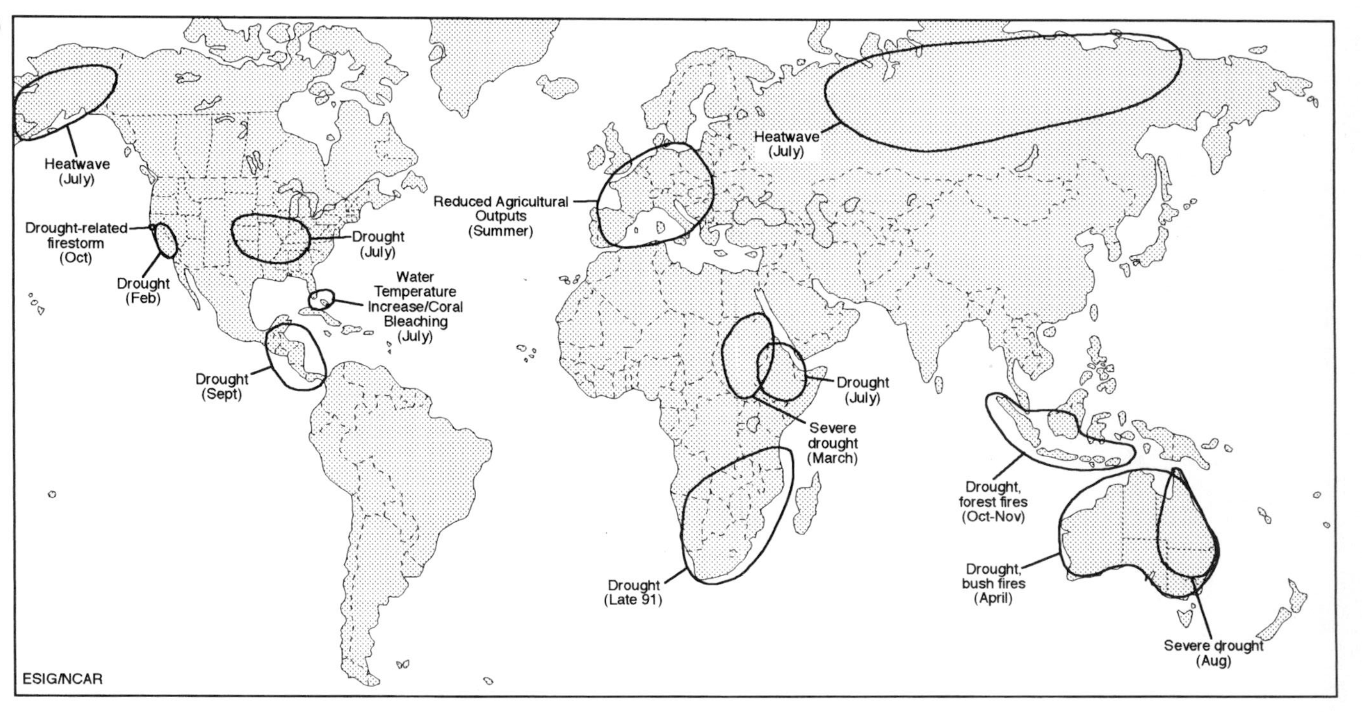
Heatwave (July)
Drought-related firestorm (Oct)
Drought (Feb)
Drought (July)
Water Temperature Increase/Coral Bleaching (July)
Drought (Sept)
Reduced Agricultural Outputs (Summer)
Heatwave (July)
Drought (July)
Severe drought (March)
Drought (Late 91)
Drought, forest fires (Oct-Nov)
Drought, bush fires (April)
Severe drought (Aug)
ESIG/NCAR

(b)

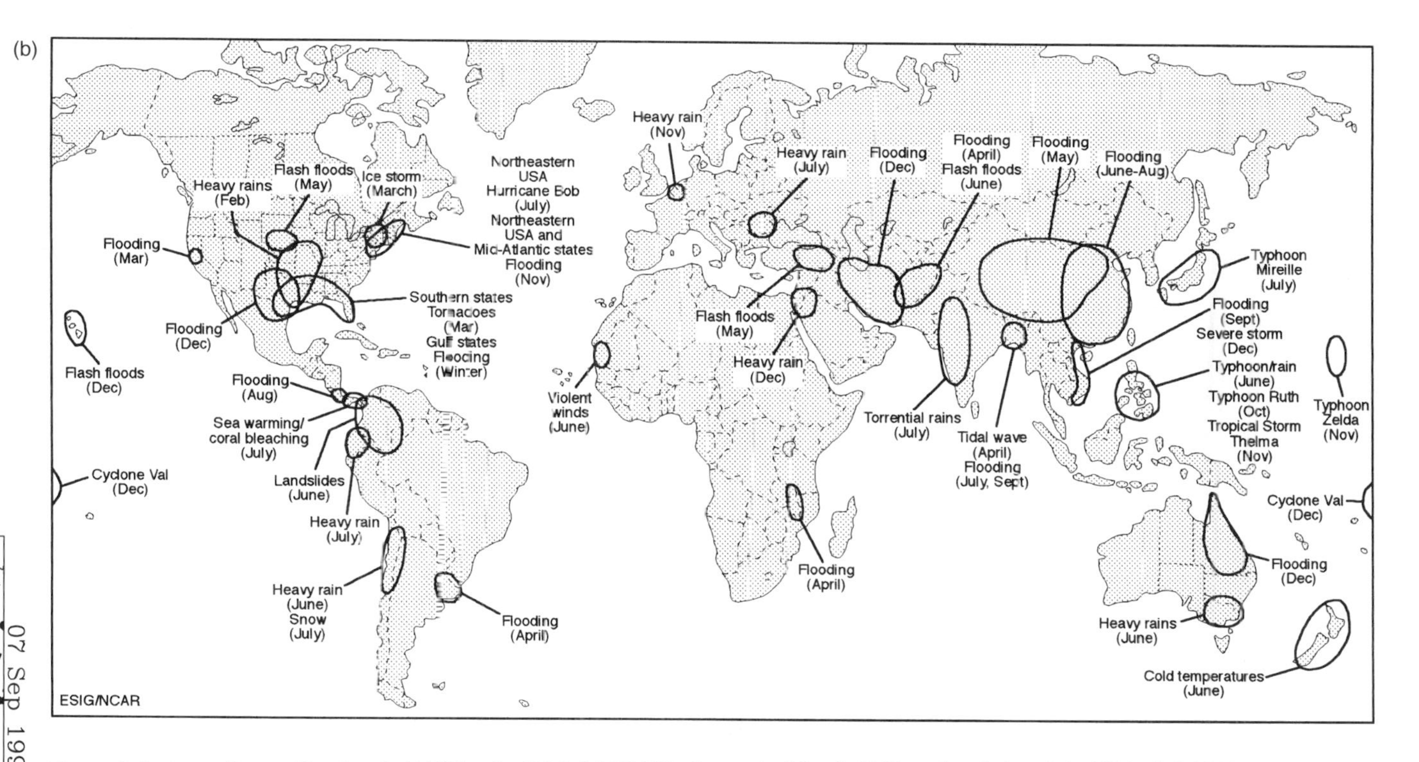


*Figure 9.5. According to Trenberth (1997), the 1991–92 El Niño began in March 1991 and ended in July 1992. (a) 1991 droughts, and (b) 1991 floods, rains, and severe storms.*

07 Sep 1997

<−1.0 −1.0 −0.5 1.0 2.5 > 2.5

(a)

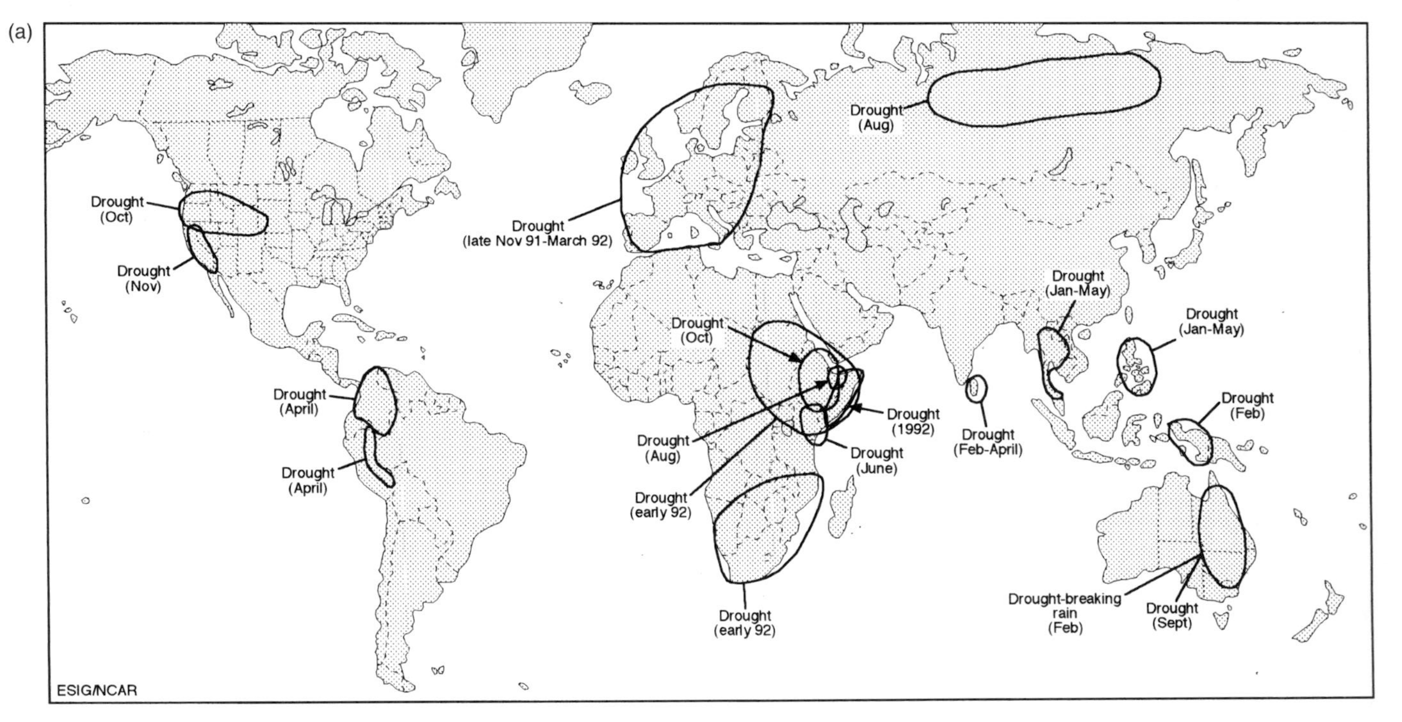

Drought (Oct)
Drought (Nov)
Drought (April)
Drought (April)
Drought (late Nov 91-March 92)
Drought (Aug)
Drought (Oct)
Drought (Aug)
Drought (early 92)
Drought (early 92)
Drought (1992)
Drought (June)
Drought (Feb-April)
Drought (Jan-May)
Drought (Jan-May)
Drought (Feb)
Drought-breaking rain (Feb)
Drought (Sept)
ESIG/NCAR

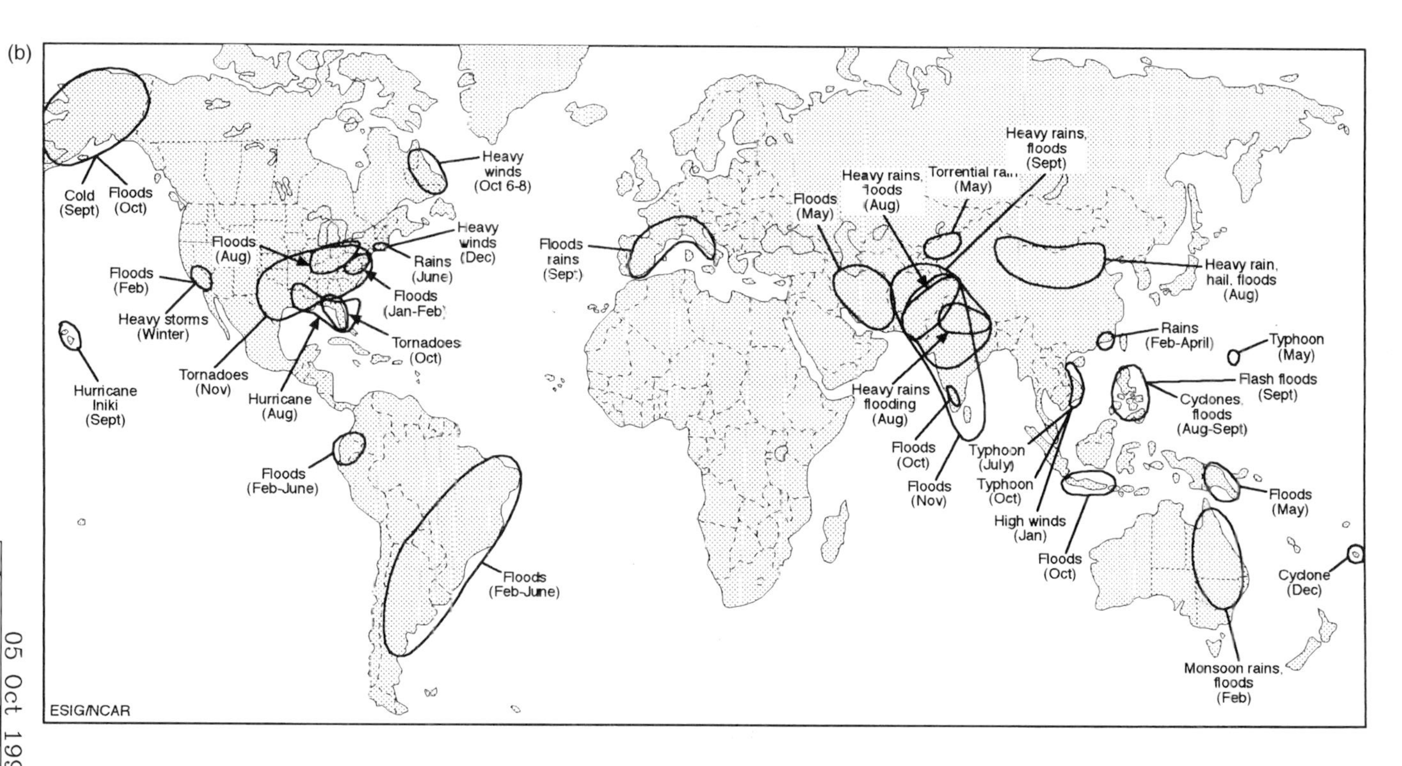


*Figure 9.6. According to Trenberth (1997), the 1991–92 El Niño began in March 1991 and ended in 1992. (a) 1992 droughts, and (b) 1992 floods, rains, and severe storms.*

05 Oct 1997

<−1.0 −1.0 −0.5 1.0 2.5 > 2.5

example, storms that usually track northward toward Alaska are shunted southward and cross the west coast of North America. Thus El Niño episodes affect both the Walker and the Hadley Circulation patterns, thereby propagating teleconnections outside the tropical region as well as within it (Figure 9.4).

Alleged teleconnections outside the tropics are more difficult to reliably attribute to El Niño, because they are further away from the central Pacific – El Niño's field of action. They are also difficult to identify because of the

## Climate change and El Niño

A proverbial dark cloud is now hovering on the horizon for the El Niño research and forecast communities. That dark cloud represents the uncertainty surrounding the possible changes in El Niño patterns and characteristics that might accompany a human-induced global climate change. Interest has been growing since the late 1970s in the potential implications of increased emissions of greenhouse gases for atmospheric processes. These gases include carbon dioxide from fossil fuel burning, methane produced from growing wet rice, nitrous oxides linked to fertilizers, and chlorofluorocarbons (CFCs) used in aerosol sprays, foam- blowing agents, and refrigerants. While the Montreal Protocol was designed to bring CFC production to an end, the Framework Convention on Climate Change is being fleshed out to reduce the global production of the remaining greenhouse gases, especially carbon dioxide.

Using large computer models, numerous climate scenarios have been derived to generate possible regional consequences of projected global warming of the atmosphere. That warming is expected to fall somewhere between 1.0 and 4.5 degrees Celsius that would probably result from a doubling of the level of atmospheric carbon dioxide and other greenhouse gases existing at the outset of the Industrial Revolution in the mid-1800s (IPCC, 1996).

Because most climate model results have described some of the possible changes of an average climate, some climate impact researchers have used such model results as a point of departure to address the possible average environmental and ecological changes that might result from future increases of greenhouse gases and to identify possible climate-related environmental changes with which future societies might have to cope (Parry *et al.*, 1988).

Impacts researchers have also pursued other lines of reasoning in order to gain a glimpse of possible societal responses to climate-related environmental changes. Few, however, have examined possible changes in extreme weather events, mainly because the model studies usually emphasize changes in average climate conditions and not changes in climate variability. Yet, El Niño events are associated with (i.e, blamed for) extreme climate-related anomalies in various regions around the globe.

various possible outcomes of the interactions among El Niño-generated perturbations and the local conditions in distant regions. *Science* journalist Richard Kerr commented on the problem facing El Niño forecasters to make consistently reliable long-range predictions for North America: "Although the equatorial teleconnection appears to be real enough, its usefulness in long-range prediction will be limited to those occasions when the signal from the tropics rises above other influences on North American weather" (Kerr, 1982). The same reasoning applies to teleconnections to other regions outside the tropics as well.

The single-event climate impacts maps for the 1991–92 event (Figures 9.5 and 9.6) provide some snapshots of the worldwide climate anomalies (droughts and floods, only some of which are truly El Niño's teleconnections) associated with that event. Remember that the sea surface temperatures only returned to average and heated up again in 1993 and again in 1994.

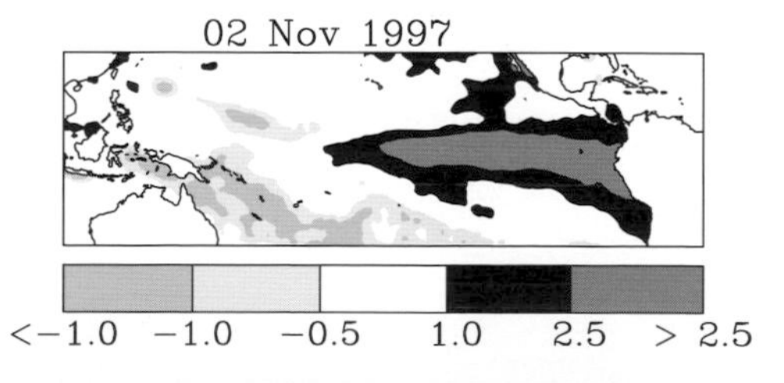

# 10 El Niño's ecological impacts: the Galápagos*

## Introduction

The Galápagos Islands, off the coast of Ecuador, are in an oceanic transition zone between the central and eastern equatorial Pacific. The islands are directly affected by both weak and strong El Niño events, whether the warm water initially forms in the central Pacific and moves toward the east coast of Ecuador and Peru, or whether it forms along the coastal area and moves toward the central Pacific.

This chapter is an abridged version of a report prepared by a researcher at the the Charles Darwin Research Station (CDRS) during the onset of the 1997–98 El Niño in the Galápagos. The report reads like a diary. The verb tenses have not been changed in order to provide the reader with the feeling of "being there" on the Galápagos during this El Niño.

## El Niño of 1982–1983 in the Galápagos

In 1982 and 1983, the phenomenon of El Niño was so forceful that scientists named it "the event of the century". Broad zones of the Ecuadorian coast suffered serious problems, which included the loss of human life and important economic losses due to floods and landslides. Galápagos was no exception. In 1983 the meteorological station of the CDRS registered 201 days with measurable precipitation, and total precipitation for 1983 was 2768.7 mm. This is compared with normal years in which measurable precipitation occurs on about 50 days and total annual precipitation averages 385 mm (based on measurements between 1965 and 1996, excluding values from 1983). Figure 1 shows the notable peak which occurred in 1983, during which El Niño reached its greatest intensity. [NB: This figure has been updated to 1998.]

The mean monthly sea surface temperature measured by the same

*Special Report from the CDRS, prepared by Michael Blimsreider (1998). Reprinted (abridged) with permission of the CDRS.
Website: http://www.darwinfoundation.org (contact cdrs@fcdarwin.org.ec).

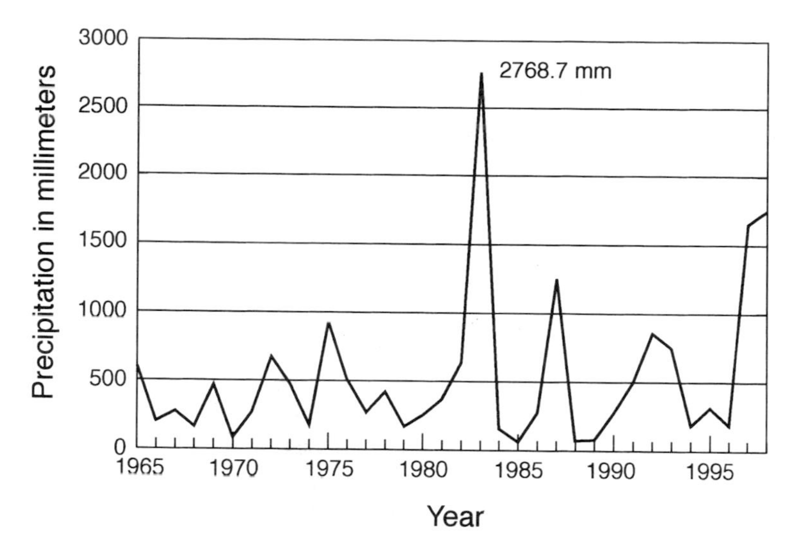


*Figure 1. Annual precipitation measured at the Charles Darwin Research Station (CDRS), 1965–98.*

station reached 28.6°C in the month of March 1983 compared to normal years in which the mean temperature of this month is 25.5°C (based on measurements between 1965 and 1996, excluding values from March 1983). Figure 2 displays mean annual sea surface temperatures observed at the CDRS between 1965 and 1996. The peak observed in 1983 is clearly visible and is the highest mean annual sea temperature observed during 31 years of observations.

Figure 3 displays changes in mean annual air temperature observed at the CDRS. Again, high temperatures peaked in 1983, when the annual mean was 25.7°C, compared with an average annual mean of 23.8°C in normal years (based on measurements between 1965 and 1996, excluding values from 1983). The highest mean air temperature observed in 1983 occurred in March, reaching 27.8°C.

The El Niño of 1982–83 brought changes that in certain cases can still be observed. The following paragraphs detail some of these changes.

- *Visitor sites:* Some visitor sites in Galápagos National Park became inaccessible due to the exuberant growth of vegetation, the destruction of paths, and the lack of safety at landing sites; additional factors such as the increase in numbers of the stinging little fire ant (*Wasmannia auropunctata*) will be discussed below. Similarly, some areas of the Galápagos archipelago possessed oceanographic

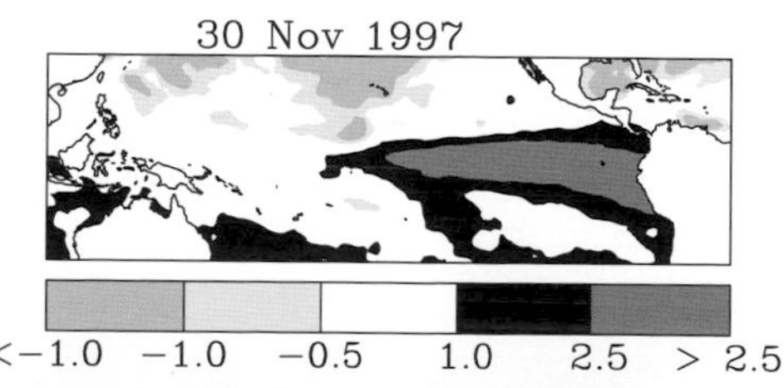

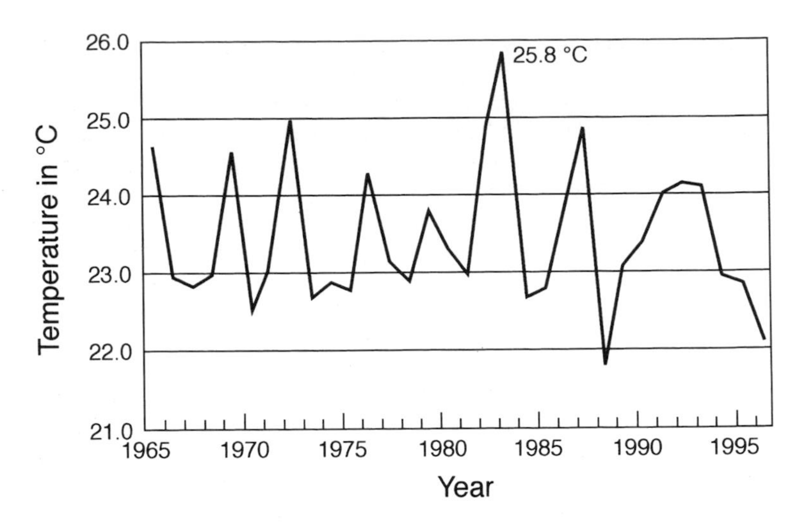


*Figure 2. Mean annual sea surface temperature measured at the CDRS, 1965–96.*

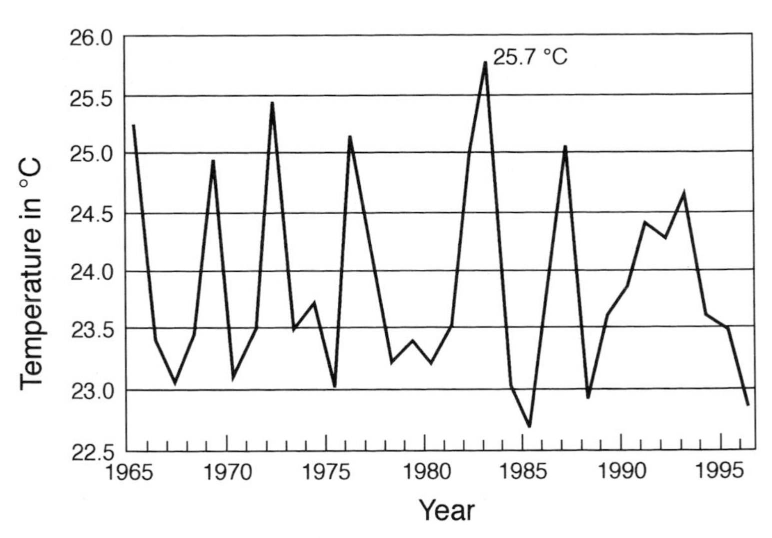


*Figure 3. Mean annual air temperature measured at CDRS, 1965–96.*

conditions that made navigation difficult, while others enjoyed unusually calm conditions that favored marine transport.

- *Fauna:* The fauna of Galápagos was affected in various ways by the El Niño event of 1982–83. Some birds and marine mammals, as well as marine iguanas, suffered visible decreases in their populations, principally due to the following factors: (a) mortality, especially caused by the absence of food and by increased incidence of illness; (b) reproductive failure, probably related to the lack of food; and (c) displacement to other locations. The effect was particularly notable in the colonies of blue-footed boobies (*Sula nebouxii*) on Española island, in the nesting of waved albatross (*Diomedea irrorata*) on the same island, in the populations of endemic flightless cormorants (*Nannopterum harrisi*) and Galápagos penguins (*Spheniscus mendiculus*), as well as in the populations of marine iguanas (*Amblyrhyncus cristatus*) throughout the archipelago. On the other hand, populations of land birds exploded; for example, the Galápagos finches reproduced in great numbers due to the abundance of food on land. [NB: A detailed account of El Niño's impacts on finches in the Galápagos can be found in Weiner's *Beak of the Finch* (Weiner, 1994).] Table 1 describes how the 1982–83 El Niño event affected the fauna of Galápagos.
- *Flora:* Immediate effects were noted in the vegetation of the Galápagos Islands, where plant populations exploded in normally arid areas and demonstrated unusually rapid and exuberant growth in the highlands of some islands. Research carried out in the months following the end of the 1982–83 El Niño event showed that the excess rain apparently had produced more effects on the abundance and distribution of various species of plants than on their numbers; the conclusions of these studies indicated that no long-term effects were expected in some species. However, there would be alterations in the type of vegetation in certain ecosystems.

Changes in the vegetation varied according to the life zone and type of plants present. Generally, species of the arid zones reacted with much greater speed to the abundance of rainfall. Herbs and climbing plants particularly took advantage of this new resource and expanded very rapidly. Similarly, enhanced germination occurred in seeds that had apparently been dormant in the soil for several months or years. Notable mortality occurred among adult individuals of the giant cacti in the genus *Opuntia*, as well as large *Scalesia* trees, which could not support their weight when their roots rotted due to excessive rainfall.

As measured by accelerated growth rates, plants in the most humid zones were less sensitive to increased precipitation than were plants in the arid zone. As in the arid zones, those humid zone species that benefited most from increased precipitation were herbs and climbers. Again, *Scalesia* was

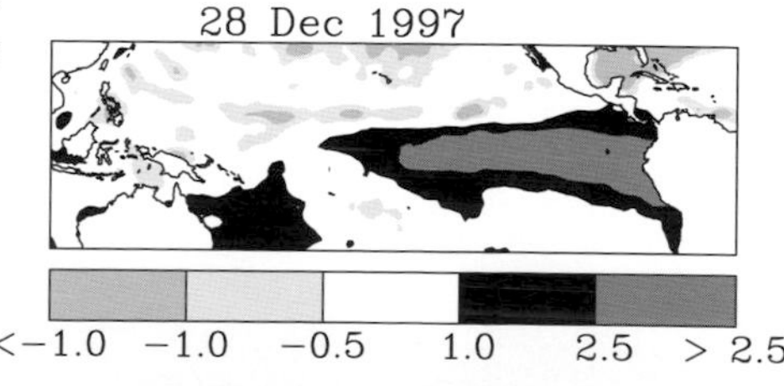

Table 1. *Effects of the 1982–83 El Niño event on the fauna of the Galápagos*

| Species | Effects of the El Niño event |
|---|---|
| *1. Marine life* | |
| Flightless cormorant (*Nannopterum harrisi*) | The flightless cormorant population suffered a 45% decrease. On Fernandina Island more than 100 individuals were found dead. At the visitor site of Punta Espinosa the birds' disappearance was obvious. By the end of 1983, reproduction had begun again and apparently had stabilized in 1984 |
| Galápagos penguin (*Spheniscus mendiculus*) | The population of Galápagos penguins suffered a 78% decrease. More than 100 dead individuals were registered in censuses, but scientists also noted a displacement of groups of penguins toward other places in the archipelago. According to the most recent Darwin Station reports, the number of penguins still has not reached the pre-event population size. For that reason, a repeat of the conditions of the 1982–83 El Niño could have serious repercussions on the current population of the Galápagos penguin |
| Waved albatross (*Diomedea irrorata*) | The waved albatross population on Española Island suffered a total reproductive failure in 1983. Of the females present, 15% tried to nest but apparently failed to hatch any chicks. The May and June rains of 1983 prevented normal incubation. About 60% of the usual nesting area was made inaccesible by vegetation overgrowth. Once conditions improved in July, scientists observed an unusual frequency of albatross mating dances |
| Great frigatebird (*Fregata minor*) | The frigatebird situation differed on each island. On Genovesa Island matings and nesting during March and April 1983 were frequent; in August of the same year, the percentage of nesting had decreased to 6.1%. On Española, however, the percentage nesting was zero, while on Pitt Islet 65% of the pairs of frigatebirds were nesting |
| Blue-footed booby (*Sula nebouxii*) | The populations of this species of booby diminished considerably. There was an elevated movement of individuals to other locations in the archipelago, but mortality was high and very obvious especially around Fernandina island. By October 1983, a recuperation in reproductive activity was observed |

| | |
|---|---|
| Marine iguanas (*Amblyrhynchus cristatus*) | Populations of marine iguanas suffered generalized and conspicuous mortality of 45 to 70% by location. Surviving marine iguanas had an average reduction of 30% in their body weight, due to the replacement of their habitual food by a species of undigestable marine algae. After the most intensive phase of the El Niño event concluded by August 1983, the populations of marine iguanas recuperated rapidly |
| Sea lions (*Zalophus californianus* and *Arctocephalus galapagoensis*) | Populations of sea lions suffered a general decrease throughout the archipelago for three principal reasons: (a) an elevated mortality among juvenile sea lions, especially from abandonment; (b) increased incidence of diseases, especially those affecting the skin; (c) migration to other locations |
| Whales and porpoises (Cetaceae) | In general, a marked absence in whales and porpoises was noted; it was assumed that they migrated to areas where food was more abundant |
| Other marine animals | Changes occurred in the composition of marine communities and abundance of various species. Abundance of some species of fish diminished, other species (typical of warmer tropical waters) appeared. There was increased and generalized mortality among corals |
| *2. Life on land* | |
| Darwin's finches (Fringillidae family) | The 1982–83 El Niño event gave rise to a considerable and conspicuous increase in the populations of various species of finches throughout the archipelago. The increase occurred because the birds nested more frequently, apparently due to the greater abundance of food. However, the El Niño rains caused many nesting failures and increased predation on chicks, especially by Galápagos mockingbirds (*Nesomimus* spp.) |
| Mockingbirds (*Conolophus* spp.) | As in the case of the finches, populations of mockingbirds increased as of 1982–83. However, in the drought years that followed, populations suffered increased mortality |
| Giant tortoises (*Geochelone elephantopus*) | The giant tortoises of Galápagos did not suffer notable population impacts. On Santa Cruz Island, giant tortoises migrated *en masse* from the highlands toward the lowlands. In the following years, measurements indicated that young tortoises grew more rapidly during the El Niño event than in normal years |

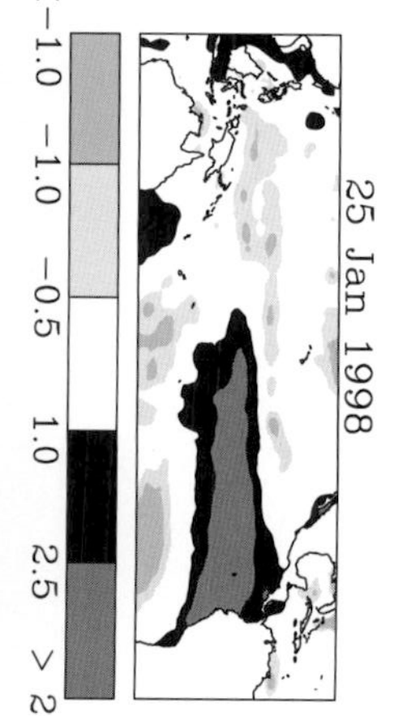

the genus that suffered greatest mortality of adult individuals (over 10 years old).

- *Introduced organisms:* During the 1982–83 El Niño event, observers noted increased abundance of introduced organisms that depend on the quantity of precipitation. Most noteworthy was the dispersal of two species of aggressive introduced ants, the little fire ants *Wasmannia auropunctata and Tetramorium bicarinatum.* The little fire ant demonstrated a particularly elevated rate of expansion, moving north on Santa Cruz island at the rate of approximately 0.5 km per year, compared with a normal rate of expansion of approximately 0.17 km per year (data taken between 1976 and 1982). The number of introduced rats (*Rattus rattus* and *Rattus norvegicus*) increased markedly in areas of the islands populated by humans.
- *Infrastructure of roads:* The infrastructure of roads was destroyed on various occasions; the highway to Baltra Island (crossing Santa Cruz Island from south to north) as well as various local roads had to be repaired continuously. The interruption of traffic caused a scarcity of goods in the highlands of the populated islands; similarly, the safety of vehicles that used the highway to Baltra was affected, especially in the section that leads to the farming village of Bellavista.
- *Generation of electricity:* Frequent power outages occurred because the generators and transmission lines were not prepared for the great quantity of rain. The unaccustomed occurrence of electrical storms was an additional problem.
- *Public health:* The general health in the populated area suffered a serious setback. According to published reports, continuous humidity resulted in an unusually high incidence of skin and gastrointestinal diseases. The heavy rains carried sediments that contaminated drinking water supplies of the population centers.
- *Supplies:* During the 1982–83 El Niño event, there was a great scarcity of various basic household staples, due both to the impossibility of reliable transport within and among the islands, and to the scarcity of the same products on the continent of South America.

## El Niño in 1997

Various global meteorological and oceanic monitoring networks detected climatic and oceanographic anomalies that began to develop in March of 1997. The first warnings were issued during the second half of the year, indicating that a phenomenon was approaching that could surpass what occurred in 1982–83. Subsequent measurements of sea surface temperatures along the equatorial portion of the Pacific displayed a growing displacement of warm water that was moving toward the western coasts of Central and South America. At mid-year the masses of warm

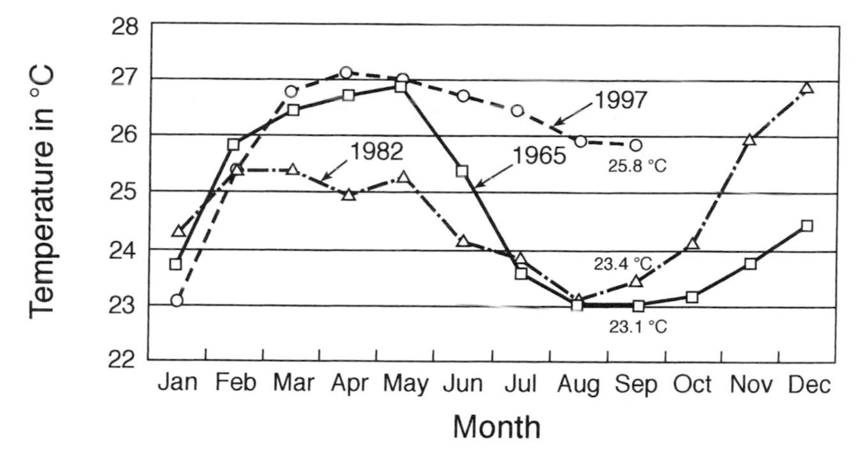


*Figure 4. Monthly sea surface temperature for the years 1965, 1982, and 1997, measured at the CDRS.*

water had reached these coasts with temperatures between 3 and 5 degrees Celsius above normal. Almost simultaneously came reports of strong rains in Chile and Peru.

In the Galápagos Islands, the second half of 1997 was marked by abnormal climatic conditions, with patterns more similar to those expected early in the year rather than in July, August, or September. Satellite imagery displayed an increase by more than 5 degrees Celsius in the sea surface temperature of the region. Similarly, the meteorological station of the CDRS registered an abnormal increase in the mean monthly sea temperature, beginning in May 1997. Figure 4 displays this increase, compared with a year that was considered to be normal (1965) and with the year previous to the El Niño event of 1982–83.

The year 1997 was characterized by strong rains during the first half of the year; Figure 5 shows total monthly precipitation through September, compared with the data for 1965 and 1982. The difference between 1997 and 1982 is noteworthy. Despite the fact that the latter preceded the El Niño event of 1983, the difference in precipitation in April of 1997 and 1982 is almost 400 mm.

The air temperature measured by the CDRS in 1997 is also various degrees above the normal temperatures observed in previous years. Similar to what is occurring in sea temperatures, 1997 is displaying patterns that differ from those which occurred in the pre-El Niño year of 1982. Figure 6 displays the mean monthly air temperatures for 1965, 1982, and 1997.

It is noteworthy that both the mean sea and air temperatures for the month of September 1997 are the highest that have been measured in the 32 years during which the Charles Darwin

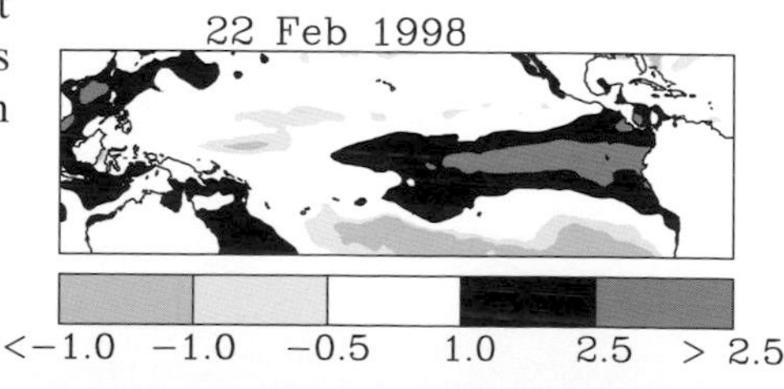

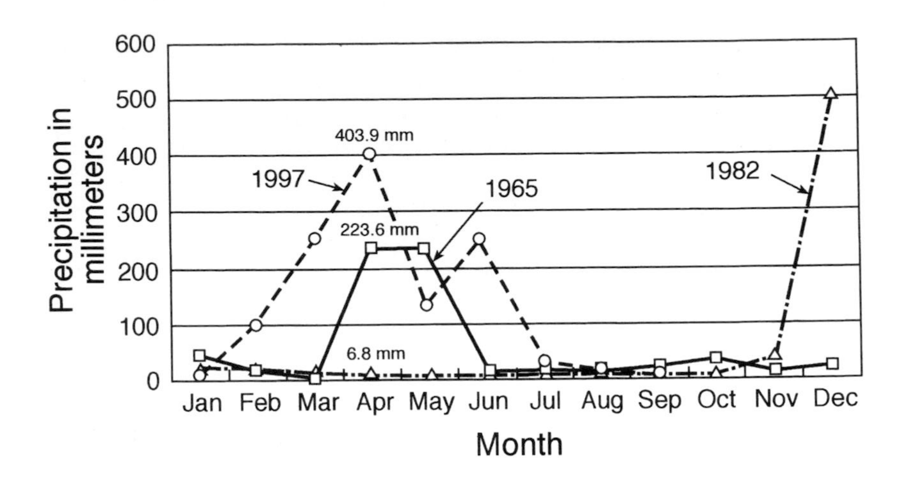


*Figure 5. Total monthly precipitation for the years 1965, 1982, and 1997, measured at the CDRS.*

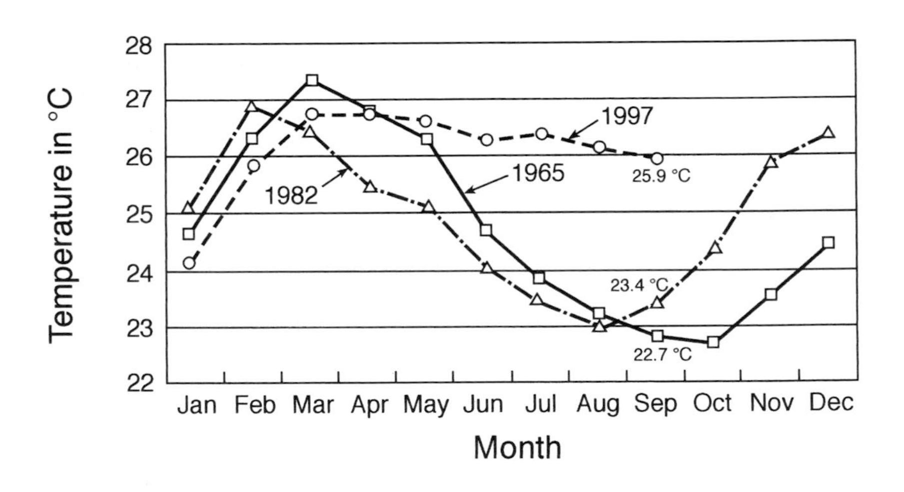


*Figure 6. Mean monthly air temperature for the years 1965, 1982, and 1997, measured at the CDRS.*

Research Station meteorological station has operated. Similarly, over the same period the April 1997 precipitation totals have been surpassed only by the rains of April 1983 (434.2 mm).

Despite the unusual conditions when compared with other years, the climate of Galápagos during the third quarter of 1997 has been relatively stable. The quantity of precipitation has diminished as of July, and temperatures have been pleasant without being exceedingly hot. Although a strong breeze predominates, many boats report that the ocean has remained calmer than is usually expected during this period in normal years ... Although no systematic study has been conducted on the fauna or flora at visitor sites of Galápagos National Park, no observations or reports have indicated any unusual mortality or changes in populations. In brief, the environmental conditions for the third quarter of 1997 are more favorable for touristic visits than in normal years.

Most current predictions indicate that the El Niño event will grow in intensity from October 1997 onward; according to recent predictions (issued in the first half of September 1997), the event will reach its peak between March and May 1998. The duration of this event is not known with any certainty.

It is noteworthy that the different climate models that have predicted the El Niño event of 1997 have been relatively accurate by comparison with predictions in previous years. This probably can be attributed to the great number of meteorological and oceanographic measuring systems that were installed throughout the world after the El Niño event of 1982–83, and to the attention that El Niño has received in recent years. The international scientific community considers that the impacts of the El Niño event of 1997 have been anticipated with greater precision than any other previous event of this magnitude.

From the point of view of economics, there is fear among the public about visiting the islands during this time. The CDRS has received a constant stream of inquiries on whether or not visitors should come to Galápagos; inquiries come both from private individuals and companies and organizations involved with tourism in the islands. According to reports of local tour operators and the Galápagos Provincial Chamber of Tourism, up to now they have received a series of cancellations from companies that do not want to risk their reputation when faced with a potentially intense El Niño event in late 1997 or early 1998. On this subject, the Darwin Station considers that, if a major event occurs, the conditions for tourism will be distinctive and may repeat the situation that occurred during the El Niño of 1982–83.

Similarly, there is concern about the physical and social effects that the El Niño event may have on the human population

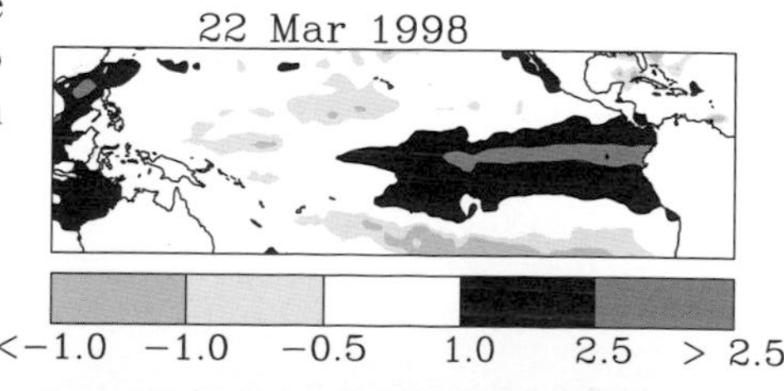

of the archipelago. In July of 1997 the CDRS and Galápagos National Park Service published a document with suggestions on how to deal with the El Niño event.

## Precautionary actions taken to date

In order to anticipate potential effects of an intense El Niño event in 1997–98, authorities have taken a series of precautions and measures, indicated in Table 2.

Furthermore, it has been proposed that Galápagos National Park and the CDRS draw up a contingency plan. This plan would include the following points:

a. production of a map of sites at risk of El Niño impacts, with different categories of risk (e.g., sites that are difficult to reach, sites to be closed);
b. a plan for intensive maintenance of sites at risk of being covered by vegetation or being destroyed by the impacts of the rain and tides;
c. an intensive monitoring plan for the sites most frequently visited;
d. alternative itineraries for tourism in case it is necessary to close one or more visitor sites;
e. a system to regularly provide information to tour operators about the status of the fauna and flora at the most heavily used visitor sites, in order to continuously update the information that tour operators can provide to passengers.

## Conclusions

Based on current analyses, the following can be concluded [as of September 1997]:

1. The Galápagos Islands are experiencing the visible effects of an El Niño event since the middle of 1997. However, although the environmental conditions are not typical for this time of year, they have remained stable during the third quarter of 1997 and are favorable for tourism.
2. Most climatic models and current predictions indicate that the El Niño event will intensify greatly during the last quarter of 1997, and will reach its maximum strength between March and May of 1998.
3. If the El Niño event reaches the intensity which has been forecasted, the Galápagos Islands will experience impacts that may possibly exceed the intensity of those experienced during the 1982–83 event. These consequences include problems of sanitation, transportation, basic services, quarantine, and social problems, as well as changes in the Galápagos ecosystems.
4. A series of precautionary urgent measures has been implemented to

counter the effects of what will probably be a very intense El Niño event.

## Latest update on the 1997–98 El Niño event in Galápagos (as of 2 January 1998)

In October of 1997, the CDRS issued a report on the situation of the El Niño event in the Galápagos Islands. After watching the development of the phenomenon during the past 2 months, we have considered it necessary to present an update to that report, in order to keep the global community informed about what is happening on the archipelago.

1. November of 1997 has definitely been the month when the rains started, as was forecast by most meteorologists.

   If we compare the rainfall figures with Figure 5 of our earlier report, we can notice that in the El Niño year of 1982 there was also a sharp increase in precipitation during the last trimester. Essentially, the rainfall and temperature now are similar to a typical rainy season, which normally would not start until around February. Conditions have been variable, with hot, sunny weather alternating with sometimes heavy (but still warm) rain.
2. The current relevant measurements at the Charles Darwin Research Station in Puerto Ayora are given in Table 3 [not shown].

   The figures for December show a sharp increase in temperature and rainfall compared with the month of November. As a matter of fact, one of the research camps located on Española Island, on the southernmost tip of the archipelago, reported 204 mm of rainfall during 13 and 14 December.
3. Reports provided by boat captains and naturalist guides indicate that the strong winds that were predominant until September have subsided. Apparently, sea conditions have generally improved, with calm, clear waters being reported all over the archipelago. However, some rough spots are encountered, especially in the northern sector of the islands.
4. Terrestrial flora and fauna appear to be flourishing with the plentiful rainfall, with abundant vegetation and food for land birds and reptiles. In southern Isabela, tortoise nesting was very limited, but in December several new nests had been reported.
5. There have been scattered, mostly verbal, reports about the current situation of the marine and marine-dependent fauna:
   - The distribution of sea lions and fur seals around the archipelago is abnormal. For example, fur seals have been spotted in places where they usually are absent, like Punta Suárez on Española Island, whereas sea lions are scarce in some of their favorite beaches.
   - The distribution of seabirds, notably boobies, has also been affected. There is no evidence of

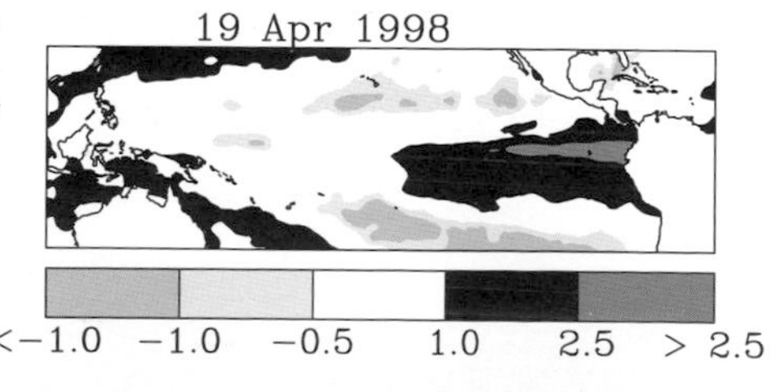

Table 2. *Precautionary actions taken in response to the El Niño event of 1997–98*

| Sector of impact | Precautionary actions taken |
| --- | --- |
| Visitor sites in Galápagos National Park | The Charles Darwin Research Station has carried out an evaluation of the infrastructure of the most heavily used visitor sites in order to provide Galápagos National Park (GNP) with a basis for establishing a contingency plan. Similarly, authorities of GNP conducted a trip to other visitor sites and evaluated potential impacts of the El Niño event on landing and disembarkment of passengers at these sites |
| Fauna and flora | The Station has produced a proposal for an environmental monitoring plan that would permit possible effects of the El Niño on the fauna and flora of the archipelago to be determined during the event. Funding is urgently being sought in order to carry out this plan |
| Introduced organisms | Continuous observations on potential dispersal events and on expansion of the range of potentially noxious introduced organisms are included within the monitoring plan indicated in the point above. Similarly, the Darwin Station and GNP have advised local authorities on immediate measures that they should take to prevent the proliferation of these organisms, especially rats and ants, in the areas populated by humans |
| Infrastructure of roads and supply lines | The infrastructure of roads on islands inhabited by humans has been radically improved. Most of the principal roads to airports and villages have been asphalted, therefore major impacts by heavy rains are not expected. Drainages have been built in high-risk sites and the highways possess systems to drain water away. There are three functioning airports and inter-island air service operates, which guarantees transport of cargo and passengers in case the environmental conditions prevent traffic by sea |
| Safety on board boats | The principal tour companies have reported that they have taken additional precautionary measures to guarantee passenger safety in case they face potentially adverse climatic conditions |

| | |
|---|---|
| Quarantine | Advances have been made to install and carry out quarantine systems and programs, both in the airports and seaports and aboard seagoing vessels, in order to avoid dispersal of potentially damaging organisms from the mainland to the islands and among the islands |
| Basic services and public health | The authorities of the Province of Galápagos have taken preventive measures to improve the systems that provide water, energy and communications. Health authorities have begun campaigns to avoid the proliferation of tropical diseases |
| Emergency procedures | The civil defense committees of the Province of Galápagos have practiced evacuation drills; the various civilian and military sectors have held frequent planning meetings to face the situation |

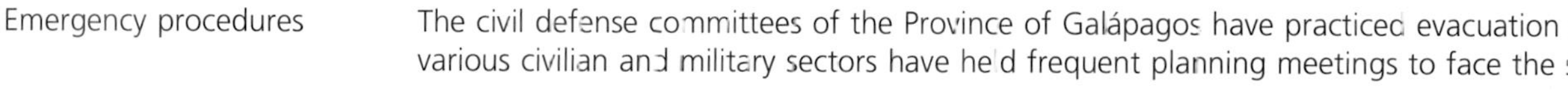

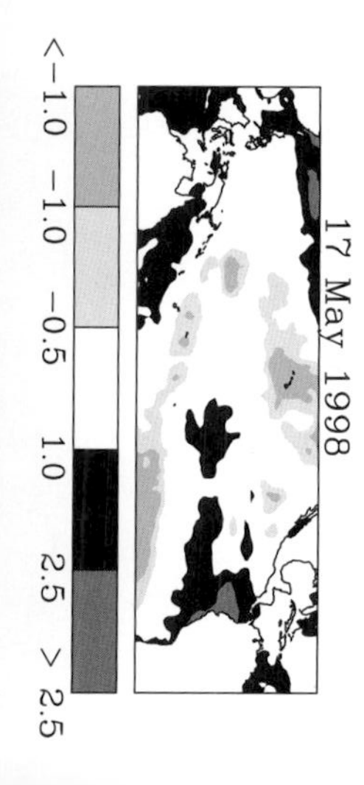

significant mortality for any species, but there has been very little successful breeding. For example, reproduction of waved albatrosses, penguins and cormorants has failed this year. Boobies have fared better.

- Hammerhead sharks have disappeared from their normal ranges, and have been found at unusual depths; this is probably due to altered water temperatures in their usual habitat. Divers have also reported changes in composition of fish population, with some species becoming more abundant and others less so, but no firm data are available.
- A recent monitoring trip to the west of the archipelago reported that green algae are diminishing in abundance, whilst red algae are increasing. This may be affecting marine iguanas; there have been recent reports of dead iguanas, slightly above normal mortality rate in certain locations.
- Sea turtles are abundant in the archipelago, as is normal for this time of the year, which is their egg-laying and mating season.

6. Effects on human settlements are already visible. Towns located in the highlands of Santa Cruz and San Cristóbal Islands have been affected by heavy rain. However, due to the improved road and drainage systems that were built during recent years, the consequences of these rains have been minor. Fortunately, the US Army Corps of Engineers was working in Galápagos as the rains set in, so a large amount of heavy equipment is available to repair any damage to the road system.

   So far, no major problems have been registered with the flights to and from the mainland (Ecuador). Delays have sometimes occurred, mainly because of problems with the weather on the mainland.
7. There have been no problems that we know of related to tourism operations. As mentioned before, the strong winds have ceased, and less wind generally means calmer seas. Passengers that have been randomly interviewed during recent weeks have generally expressed their satisfaction with their trips, noting that it has been raining but that this is a minor inconvenience. The clear, warm waters have been favorable for snorkeling.

## Revised conclusions and projections

The Station issued four conclusions in its past report. Below, we are updating these statements in the light of the new data.

- The Galápagos Islands are experiencing the visible effects of an El Niño event since the middle of 1997. The environmental conditions are not typical for this time of the year, and the stable conditions observed during the third quarter of 1997 have been followed by the onset of hot, rainy weather. So far, the ecological consequences have been most obvious on land with the abundant plant life. Marine flora and fauna are also

Table 3. *Puerto Ayora measurements at CDRS, September–December 1997*

| Month (1997) | Air temperature (in °C) | Sea temperature (in °C) | Rainfall (mm) |
|---|---|---|---|
| Sep. | 25.9 | 25.8 | 5.8 |
| Oct. | 25.9 | 26 | 7.9 |
| Nov. | 26.5 | 26.7 | 146.9 |
| Dec. | 27.2 | 28 | 317.5 |

changing in abundance and distribution. Hard data on the ecological effects are scarce; we have to rely largely on individual observations. However, we consider that current conditions for tourism, although uncomfortable due to the increased rain, still allow for a unique wildlife experience. For visitors interested in ecology, the chance to observe first-hand the effects of El Niño may itself be of interest.

- Most meteorologists predict that the hot, rainy conditions may intensify in the first part of 1998 and will be maintained through to April or May. The CDRS will continue to measure rainfall and temperature but is not in a position to confirm or reject the meteorologists' forecasts.
- If the El Niño event reaches the intensity and duration that have been forecast, the Galápagos Islands are likely to experience intense ecological impacts. These impacts favor some species and are detrimental to others. The experience of the 1982–83 El Niño indicates that animals and birds dependent on the sea will be the worst affected. However, it should be cautioned that the history of this year's event has not been identical to that of 1982–83, so the consequences may also be different.
- From the point of view of long-term conservation, the greatest concern is that El Niño will facilitate the establishment and spread of species alien to Galápagos, causing permanent ecological change.
- In addition to these changes in the ecosystem, the El Niño of 1982–83 caused problems for the local communities, with regard to sanitation, transportation and basic services. This year it seems that improvements in the overall conditions of human settlements and road systems have prevented the onset of major problems. Further observation is necessary, though, if rainfall keeps increasing as predicted. Most tourism takes place out in the uninhabited islands, so it is not affected by these problems.
- A series of precautionary measures, described in the earlier report, has been implemented to counter the effects of what apparently is becoming a very intense El Niño event. So far, we consider that these measures have been successful.

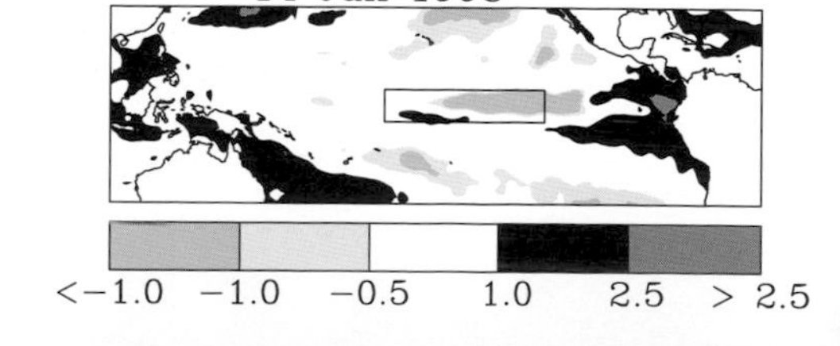

The Charles Darwin Research Station is urgently seeking funds to monitor El Niño events and related ecological changes, and to be able to respond to problems such as the introduction of new alien species. We would also welcome approaches from experienced, self-funded scientists wishing to propose El Niño-related research projects. This is an important time for Galápagos conservation and an opportunity for scientists to learn more about the El Niño phenomenon, for both local and global benefit.

# 11 Methods used to identify El Niño

Scientists have used a variety of methods to gather and analyze information about El Niño and La Niña. Such information is used by some researchers to improve the understanding of the El Niño process and by others to forecast its worldwide impacts. Some methods are focused on making direct observations of certain key characteristics of the warm event/cold event cycle. Other methods have relied on indirect (proxy) information, including pollen samples, ice cores, and soil layers, etc. This information is used to identify, for example, the impacts on the land's surface of floods, droughts, or fires that are perceived to be linked to El Niño. Each of these natural hazards leaves its unique mark on the natural environment. A careful assessment of soil profiles, for example, can identify some of the climate-related impacts that may be attributed to El Niño (e.g., floods in normally arid lands such as coastal Peru or Chile).

Other research tools and methods include remote sensing from satellites, statistical assessments, historical studies, and computer modeling and simulations. While the results of the various approaches may not agree with one another, for the most part, each scientific study, each new statistical method of assessment, and each new monitoring scheme adds to the scientific understanding of the El Niño phenomenon. Even negative findings can be viewed in a positive light, because they identify "blind" research alleys to be avoided by others in the future.

Before the age of satellites and computer models, researchers interested in the Pacific Ocean (or any other ocean, for that matter) had no choice but to rely on research cruises and ships of opportunity; that is, commercial ships that criss-cross various parts of the world's oceans. These ships would record *in situ* information about ocean currents, surface winds, sea surface temperatures, and so forth. However, the Pacific Ocean is a very large body of water, and the necessary reliance of researchers on data gathered during scientific cruises and on "opportunistic" observations of ocean conditions by transoceanic vessels left large gaps in the data coverage of the Pacific Ocean surface in both space and time for a variety of environmental changes that could be related to El Niño.

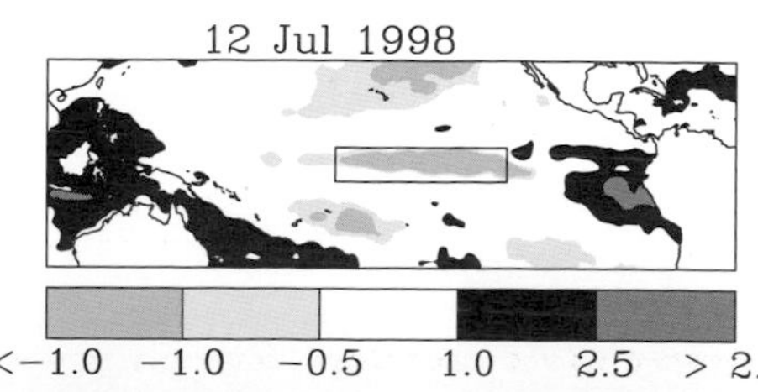

## Statistical methods

Given that relatively few El Niño events have been well observed by the current working generation of researchers, statistics play a major role in research on El Niño. The use of statistical methods (such as correlation and multiple regression) was pioneered by Sir Gilbert Walker in the earliest decades of the twentieth century. Walker also relied on statistical techniques (especially time-series analysis) to identify and model the recurring features of the Southern Oscillation. His methods, as noted earlier, helped to identify many of the worldwide teleconnections associated with the occurrence of the atmospheric pressure changes in the tropical Pacific, i.e., the Southern Oscillation phenomenon. The use of statistical methods has also enabled modelers who study the general circulation of the atmosphere and of the ocean to incorporate all kinds of relevant physical processes into the modeling and forecasting of El Niño. Current long-lead forecasts of climate anomalies, such as those issued by the US National Weather Service's Climate Prediction Center, combine the outputs from coupled atmosphere–ocean general circulation models with multivariate statistical techniques (such as canonical correlation analysis or principal component analysis) in order to produce El Niño and La Niña forecasts with a lead time of 3–6 months and longer.

The importance of the use of statistics in El Niño-related research cannot be overstated. Statistical assessments pervade all aspects of such research from reconstructing paleoecological records to identifying potential teleconnections to projecting the possible impacts of a global warming of the atmosphere on various characteristics of El Niño. They have been a necessary and integral part of using El Niño information, for instance in developing a mechanism for forecasting hurricane frequency on a seasonal basis (e.g., through the use of contingency tables).

For their part, social scientists, along with other potential users of El Niño information such as agronomists, hydrologists, and fisheries managers, are also very dependent on the appropriate use of statistics to identify the probable locations and strengths of the impacts of El Niño episodes. With such information they can advise governments and citizens more effectively on how they might best adapt to, mitigate, or prevent their potential adverse consequences for human activities.

## In history

The public tends to think that much of our knowledge about El Niño has emerged only in the last decade or so of the twentieth century. However, as noted earlier, interest in and research on aspects of El Niño can easily be traced back at least to the late 1800s. The major El Niño in

1891 sparked local discussion among Peruvian geographers of El Niño as a phenomenon that they should be concerned about. Following that event, Peruvian navy captain Camilo Carrillo described the history of scientific interest in Peru's coastal waters, going back to the late 1700s and 1800s. His knowledge about El Niño suggests that Peruvians had a great deal of information in hand by the end of the nineteenth century. He spoke of changes in coastal winds and in biological productivity, along with awareness of excessive rainfall and flooding in northern Peru and disease outbreaks associated with what we have come to know as El Niño-related changes in the ocean environment. So today, much of the El Niño-related information related to Peru that today's scientists are using as their base of information must have been known to some extent for about a hundred years, even though what they observed was not called El Niño as far as we now can tell.

### *Anecdotal information*

Anecdotal evidence appears in a variety of forms: in personal diaries and in historical interpretations of past events. For example, in 1895 an American geologist, Alfred Sears, gave a lecture to the American Geographical Society on "The Coastal Desert of Peru". In addition to a brief description of the flora, fauna, and geology of the coastal region, he referred briefly to the "septennial rains" that occurred in the usually arid northern Peru. These rains seem to have occurred every 7 years or so and have since been viewed, with good reason, as the regional rains that usually accompany El Niño events.

Sears wrote about the occasional great changes that he observed in the desert landscape: "the hitherto lifeless earth springs into being; gross and flowering plants appear on every hand, grown to the height of a horse's head" (Sears, 1895, p. 262). He also recounted local folk history. For example, in 1531, when the Spanish conquistador Francisco Pizarro began his conquest of Peru, the septennial rains proved to be his ally. He had apparently landed with his troops along the northern Peruvian coast during a wet (read this now as El Niño) year. This enabled him to find food and water for his men and their horses along the arid route of his conquest, food and water that in most years would not have been available. Sears described it in the following way:

> It rains on the northern coast of Peru only once in seven years. All the remaining years are utterly dry. Pizarro could not have gone from the Tumbez [northern Peru], where he first landed after his fight with the Indians at Puna, to the valley of

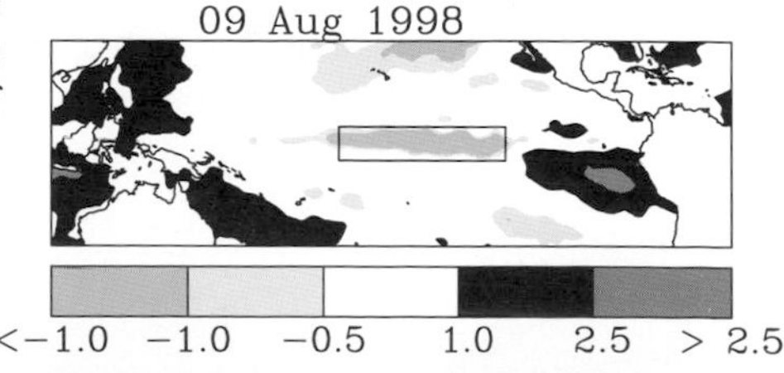

> Tangarara and found feed for his animals, nor would he have found the little settlements mentioned as existing along the road in any other season than a wet one. Again, he would not have been driven from the Chira Valley, Tangarara, by malarial fevers, save in a wet year, as it exists in that valley only in wet years.
>
> (Sears, 1895, p. 263–54)

Interestingly, in 1985, 90 years later, several articles appeared in *Disasters*, an international journal of disaster studies and practice, on the impacts of the 1982–83 El Niño on human health in some South American countries. One of the papers discussed the research finding that outbreaks of malaria (as well as of dysentery and cholera) tended to accompany El Niño events. A decade later in early 1995, newspaper articles once again described recent research findings linking El Niño and adverse health effects. By 1999, the health effects of El Niño had become one of the new and exciting key areas of government interest about infectious diseases (malaria, dengue, Rift Valley fever, etc.) (Epstein, 1999; Colwell and Patz, 1998; see also Chapter 12).

Yet Sears, more than a hundred years earlier, had uncovered the local belief in the connection between these septennial rains (a proxy for an El Niño event) and malaria outbreaks in northern Peru. Such historical anecdotal comments are apparently being "rediscovered" today by researchers in a variety of fields. Researchers are placing an increasing value on El Niño-related anecdotal information for use in research.

American oceanographer William Quinn and his colleagues relied on the use of all kinds of historical information that they considered to be related to El Niño events. Anything that provided clues about the possible occurrences of El Niño was considered. As a result, their chronology was drawn from personal diaries, ship's logs, rainfall records, plantation data, and so forth (Quinn *et al.*, 1987). Later, this chronology was extended using more distant locations of alleged impacts of El Niño, such as Nile floods, with records going back to the seventh century. Thus researchers have constructed a chronological history of El Niño episodes for the past several hundred years, based on proxy record.

### *Paleoecological evidence*

Referred to as paleo studies, several efforts to reconstruct El Niño's history well back into prehistoric times seek to identify indirect signs of the occurrence, frequency, and magnitude of El Niño events. Some researchers have even tried to identify more specific characteristics of those events using proxy data, as derived from analyses of tree rings, ice cores, fossil soil deposits, marine sediment records of fish abundance, evidence of widespread fires, floods, droughts, and so forth. These types of data provide indirect indications of the occurrence and, in some cases, the intensity of past El Niño episodes, their local environmental impacts, and the existence of possible teleconnections (see Diaz and Markgraf, 1992).

Clearly, proxy information has been very instrumental in identifying changes over long periods of time in the various characteristics of El Niño events. However, there are several assumptions that need to be proven: (a) that El Niño events of the distant prehistoric past were similar to those of today; (b) that their patterns of teleconnections, as we know them today, would have been similar under the different atmospheric and oceanic conditions that existed then; and (c) that El Niño processes responded to the same forcing conditions, despite centuries-long alternating periods of warm and cold global temperatures. (For example, would a human-induced global warming evoke ecological changes similar to those that would accompany a naturally occurring global warming of the same magnitude?) There is also a problem with intervening events such as periods of eruptions of active volcanoes around the globe, the impacts of which could either hide the occurrence of an El Niño event or could produce effects similar to it (i.e., mimic El Niño) but in the absence of an actual El Niño event. This generates the problem of attribution; that is, trying to identify cause-and-effect relationships when there may have been many other factors contributing to an environmental change.

These words of caution notwithstanding, reconstructing El Niño's past history has been extremely fruitful and has led to an improved understanding of its causes and effects and of changes over time in its characteristics and impacts.

### *Biological evidence*

The 1957–58 El Niño and its ecological impacts captured the attention of scientists at a California symposium in 1959. At the meeting, participants highlighted the importance of biological indicators of changes in the ocean, suggesting that "It was abundantly evident ... that the strongest and most spectacular evidence of marked change in the coastal countercurrent ... came from biological, rather than physical, observation" (CalCOFI, 1959, p. 211). Early changes in oceanic conditions contributing to El Niño's onset may have had seemingly negligible (but nevertheless detectable) impacts on some living marine resources well before other more easily measured physical indicators had become evident.

Several biological indicators (actual and potential) can be used to identify the onset of an El Niño event. One such indicator is the age structure of fish catches. For example, during the 1972–73 event, pockets of cold water appeared close to the Peruvian

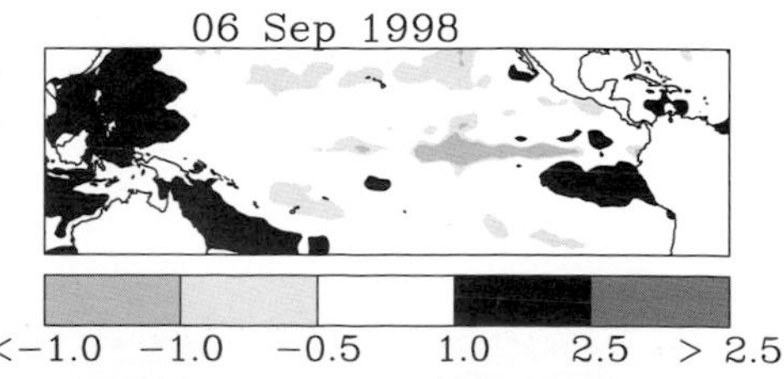

coastline in the midst of a general sea surface warming along the Peruvian coast. These pockets contained large numbers of anchoveta that were easily caught by fishermen. The fishermen pulled them into their boats at very high catch rates and ferried their catches back to the fishmeal-processing factories along the coast, only to return later for additional catches (sometimes in the same day). Unfortunately, fishermen were not looking closely at changes in the age structure of their catches; adult anchoveta were being caught in extraordinarily high numbers. Yet, this portion of the fish population was needed to produce future generations. By the time fishermen and researchers realized what was happening, the anchoveta population was in the midst of a collapse from which it would take more than a decade to recover.

Another biological indicator of El Niño relates to the fat content of a fish. During El Niño events, the fat content is reduced as the fish draw on their fat reserves for sustenance in the presence of nutrient-poor surface waters. Linked to this is yet another biological indicator of changing environmental conditions, the Gonadic Index. Chilean researchers used the Gonadic Index in the 1982–83 event to identify a decrease in the fertility (in this instance) of the Chilean sardine (Serra, 1987), and they continue to this day to do so. When these fish have low fat content, little fat is available for gonadic development and, consequently, for egg production, which causes a decrease in overall fertility. Thus changes in fat content can be used as proxy information to identify subtle ecological changes in the marine environment in the tropical Pacific, and perhaps can be used even as input to the various El Niño early warning systems.

Yet another biological indicator of El Niño-related changes is the composition of fish species in Peru's coastal waters. When warm water appears off the coast, cold water species disappear and warm water species appear. As a result, catches of warm water species by fishermen increase during El Niño. For example, wild shrimp become prevalent along the coast of southern Ecuador and northern Peru during El Niño. In this regard some Peruvian coastal communities benefit from El Niño conditions along the coast (Sharma, 1999). Those who benefit, however, are not the same as those who suffer. In general, fishing gear is designed for capturing certain species and, to capture different species, fishing boats must be fitted with different gear – a costly operation.

The behavior of the bird population has also been used as an important indicator of environmental changes related to air–sea interaction in the tropical Pacific Ocean. During El Niño, bird populations on various islands in the tropical Pacific, such as Christmas Island and the Galápagos Islands, abandon their colonies, fail to reproduce, and suffer high mortality rates, especially among the chicks. For some bird species, this is related to the decrease in the abundance or in the availability of local fish populations on which the birds feed. During intense warm events, nutrient availability declines sharply, fish reproduction fails and the remaining fish population disperses, making fish more difficult for birds to locate. In fact, satellite pictures clearly show the reduction in the nutrients near the ocean surface around such islands and coastal areas. This was a prominent feature during the 1982–83 and the 1997–98 El Niño events (see Chapter 10) and has been linked to the observed reproduction failure of seabirds and certain marine life around the Galápagos Islands.

> A recent investigation, utilizing satellite ocean color observations, and complemented with coincident oceanographic measures, has demonstrated the tight coupling that exists between the distribution of phytoplankton populations around the Galapagos Islands and the oceanographic conditions observed during the 1982–83 El Niño.
>
> (Feldman, 1985, p. 125)

This relationship was exhibited again in the 1997–98 event by way of satellite imagery. Imagery identified a large expanse of increased biological productivity in the ocean in the area of the Galápagos, providing researchers with an early indicator of the end of El Niño conditions in the region (Busalacchi, 1998). Biological productivity is a result of upwelling processes, and the increase in such productivity indicated a return of upwelling processes near the equator that occur under non-El Niño conditions.

## Observations, monitoring, and modeling

Observations of ocean temperatures from the surface to the greater depths is critical. Such measurements on a continual basis are needed by those responsible for forecasting climatic conditions around the globe. Also, researchers need this information to calibrate their model's parameters as well as their model's output and their analyses against the physical reality. Remote sensing specialists use *in situ* observations and measurements for the development and refinement of their remote sensing tools.

### *Remote sensing*

With the advent of satellites, the potential for improving society's ability to monitor, forecast and respond to the occurrences of El Niño and La Niña events has greatly increased. Various parameters can be monitored from space over time across large expanses of the oceans. NASA researchers K.-M. Lau and Antonio Busalacchi (1993) have suggested that

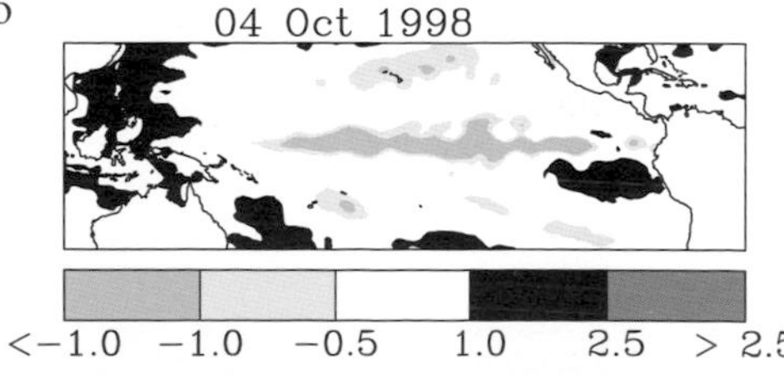

> the picture of a coherent, global-scale variation [i.e., El Niño] did not emerge until the late 1960s and early 1970s when weather satellites began to appear on the scene. In fact . . . [Bjerknes'] . . . conclusion of large-scale, coherent variations across the Pacific was largely based on composite satellite cloudiness pictures from [satellites] ESSA3 and 5.
>
> (Lau and Busalacchi, 1993, p. 281)

Since the mid-1980s, El Niño events have been closely monitored at the Earth's surface, within the ocean, and from space. However, the 1997–98 El Niño was the first event to have been comprehensively monitored, because of the completion of the TOGA-TAO Array of buoys in the equatorial Pacific in 1995. Lau and Busalacchi (1993) noted the degree of success of the problems confronted by a growing dependence on the use of satellite remote sensing of the following key El Niño parameters: sea surface temperatures, latent heat and moisture fluxes (this relates to rain-producing atmospheric processes and to the atmospheric cooling or warming of the ocean), atmospheric columnar water vapor (this relates to the rain potential of the atmosphere and the dynamics of the ocean and atmosphere), surface wind stress (winds that drag the upper layers of the ocean), ocean circulation, and changes in sea level. Extolling the value of remote sensing, they concluded that

> Because of the close coupling of these atmospheric and oceanic variables, the simultaneous and coordinated monitoring of the entire ocean–atmosphere system by different satellite instruments is paramount.
>
> (Lau and Busalacchi, 1993, p. 291)

Satellites enable researchers to observe the results of numerous interacting environmental processes without having to observe each and every subprocess in isolation. One good example – among many – of the potential "power" of remote sensing as an El Niño monitoring tool is its ability to measure thunderstorm activity indirectly by monitoring outgoing longwave radiation (OLR). Briefly stated, the incoming energy from the sun (solar radiation) in the form of *shortwave* radiation reaching the Earth's surface is re-emitted to space as outgoing *longwave* radiation. OLR can be intercepted (blocked) by cumulus cloud systems, which are good candidates for producing rain. Thus, in regions where there is such cloud formation, OLR readings are lower than where clouds are not present to block the outgoing longwave radiation. The shift of rain-producing convective activity toward the central and eastern Pacific at the onset of an El Niño episode can be captured by satellites that monitor OLR.

## Modeling activities

Information about the Pacific Ocean's behavior has come from measurements from stationary and drifting buoys and from ships on their way from one location to another. And, although the satellite era has ushered in new tools and new ways to measure various aspects of the oceanic and atmospheric environments, neither the old nor the new monitoring tools can be relied on totally to provide all the pieces to the El Niño puzzle. The old methods of data collection would take too long and are too costly for gathering information over such a large expanse of ocean and across the decades and centuries in order to monitor in detail a statistically significant number of El Niño events. Because of these limitations, a growing number of researchers believe that future researchers will have to rely to a greater extent on computer-run general circulation models in order to gain insight into El Niño processes for forecasting and for other scientific purposes. (In the meantime, however, statistical models seemed to have fared slightly better in forecasting the 1997–98 El Niño than the dynamical general circulation models (Barnston *et al.*, 1999.)

General circulation models are by definition simplifications of reality. Given their current state of knowledge, scientists have developed mathematical expressions for their models that represent what they consider to be key aspects of the ocean and the atmosphere, and their interactions. The models are then tested to see whether they are able to reproduce the occurrence of past El Niño events. Once satisfied that these models can reproduce the precursor stage of an El Niño with some degree of reliability, modelers then turn their efforts and models toward (a) projecting the onset of future El Niño events, or (b) attempting to understand how the atmosphere and the ocean interact to produce the El Niño phenomenon, or (c) identifying teleconnections in tropical and extratropical regions.

There are different types of model, from the simple to the very complex. No single model can capture all aspects of the observed El Niño or La Niña events. In general, the most widely used models are called coupled ocean–atmosphere models. These types of model are used to replicate the dynamic interactions between oceanic and atmospheric processes and fall into three main classes:

- *Limited-area models.* This type of model focuses on a relatively small geographic region within the Pacific Ocean and on a corresponding limited part of the atmosphere, rather than on either the entire globe or the entire Pacific basin. Typically, the areal coverage of this model is the tropical part of the Pacific basin. These computer models are relatively simple in their mathematical formulation. They are easy to run on modern microcomputers, and are relatively economical. A limited-area model was used by oceanographers Mark Cane and Stephen Zebiak in 1986 to successfully forecast the 1986–87 El Niño. Until the most recent El Niño in 1997–98, several researchers saw the Cane–Zebiak model as a good candidate for making reliable fore-

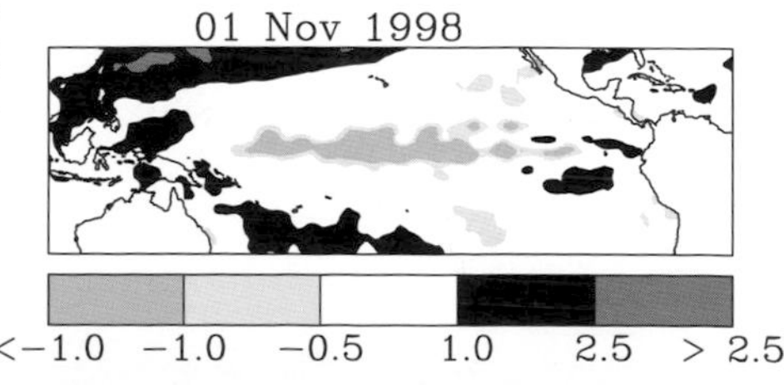

casts of El Niño on an operational basis. Their model, however, not only failed to forecast the onset of the 1997–98 event but, as noted earlier, had called for a La Niña instead! It has since been reviewed and modified to improve its projections. However, its status as the flagship model to be used for forecasting El Niño was diminished.

A number of investigators have used limited-area models to investigate the development and effects of internal ocean waves (i.e., Kelvin waves and Rossby waves) in producing either El Niño or La Niña events. This type of model has been criticized because its relative simplicity and its limited domain (i.e., the geographic area of coverage) exclude processes that may prove to be important for understanding the El Niño phenomenon and the possible influences on the tropics of regions outside the tropics.

- *Global atmospheric general circulation models* (GCMs). One type of a global atmospheric circulation model is linked by mathematical formulas to a limited-area ocean model. Typically, the limited-area ocean model is highly detailed and capable of correctly reproducing internal ocean wave dynamics.

  Because the ocean part of the model has a limited geographic scope, there are often problems in capturing oceanic processes that occur outside the model's geographic domain, particularly those processes in the western Pacific Ocean or in other oceans.
- *Coupled general circulation models*. Another type of global atmospheric GCM is one that is linked to a global ocean GCM. These coupled models are particularly useful for generating global warming or cooling scenarios and for identifying their possible impacts on some aspects of an El Niño, such as changes in its intensity, magnitude, and duration.

  However, these models are very costly to run, requiring large amounts of time on supercomputers. So, scientific researchers have to make tradeoffs between a model's mathematical complexity (which may better represent the real world) and cost (of running relatively simpler models for shorter periods of time).

No single type of model is capable, by itself, of capturing all aspects of El Niño events. As most of these computer modeling activities are still relatively recent undertakings, and as each of the models has significant limitations, their use, application and/or value for forecasting or for societal impacts research purposes is debated within the scientific community (e.g., Broad 1999; Pfaff *et al.*, 1999). In particular, there is widespread discussion by modelers and policymakers alike about how best to use the models to make seasonal or long-lead El Niño forecasts on an operational basis or to provide estimates of how the El Niño process might be affected by global warming. Computer models, as research tools, will probably continue to improve during the next decade in their ability to model El Niño-related processes, and possibly improve forecast reliability.

Thus modeling results related to seasonal to interannual forecasts must be viewed in the context of the errors contained within the present-day models. This underscores the fact that great care should be taken by those who seek to use any specific El Niño model's output for decisionmaking purposes. The need for such care was again reinforced by the results of an assessment of how well 15 El Niño modeling activities fared in forecasting the onset of the 1997–98 event (see Chapter 8).

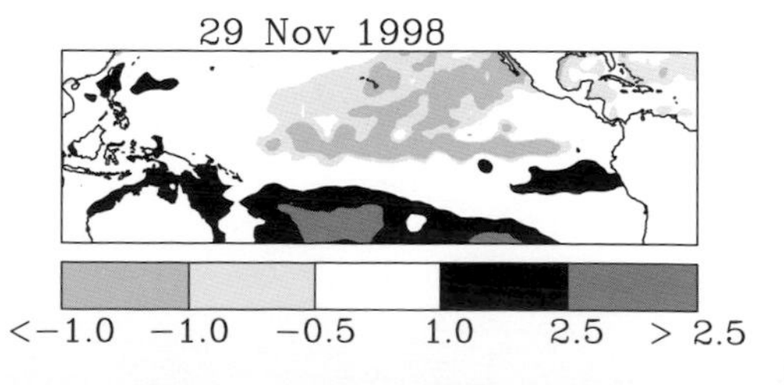

Section III

# Why care about El Niño and La Niña?

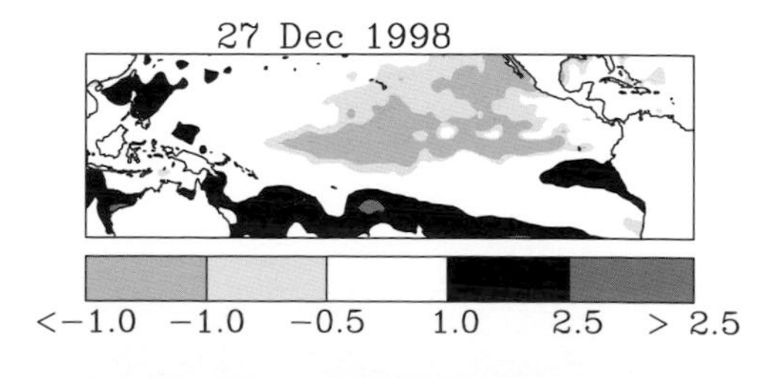

# 12 El Niño and health

## Historical interest in climate and health

Throughout history, societies have been interested in climate's impact on public health. Although just about everyone in society at one time or another has been concerned with the adverse effects of those impacts, for the most part it has been a topic left for health specialists (broadly defined) to deal with. Those who are in various aspects of the medical profession have been given the responsibility to identify factors that increase health-related problems and then to develop ways to cope with them. Interest in infectious diseases seems to increase as it worsens from the usual (i.e., expected) levels of its seasonal occurrence to outbreaks and epidemics. Epidemics, like those that accompany famines, provide graphic examples of the ever-present vulnerabilities of populations around the globe to disease. Pandemics seem to attract the most attention, because they involve large numbers of people, encompass large areas, and result in a large number of fatalities.

Despite the high levels of technological advancement that many countries had achieved by the end of the twentieth century, societies everywhere – rich and poor, inside and outside the tropics, in cold environments and hot ones, in dry areas and in wet ones – are still highly vulnerable to infectious disease outbreaks (World Resources Institute, 1998). With the high level of population mobility, numerous military conflicts, changing environmental conditions, increasing poverty around the globe, and globalization processes in general, the ease and rapidity with which such diseases can spread have become phenomenal as we enter the new millennium.

Present-day scenarios about the effects of climate extremes and climate change on the spread of infectious disease seem to border on science fiction. In fact, major Hollywood movies have capitalized on the public's fascination with the threat of epidemics. Under such movie titles as *Contagion*, *Outbreak*, *Virus*, etc., the US film industry has

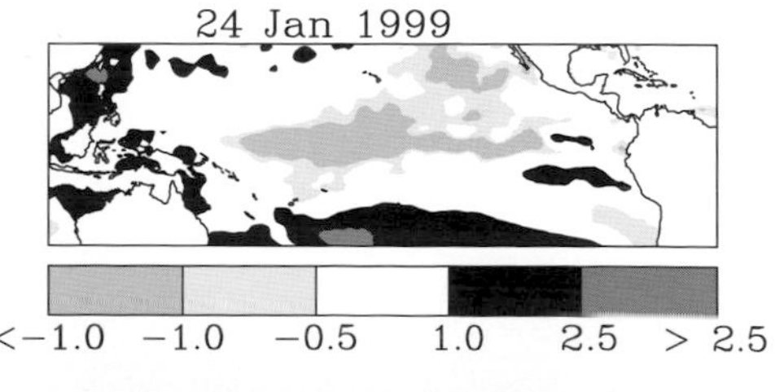

focused on the precarious nature of public health and turned medical fiction into feature films.

The recent ebola outbreaks in central Africa (described in a book by Richard Preston (1995) entitled *The Hot Zone*), dengue hemorrhagic fever outbreaks in Vietnam and in Texas, hantavirus among American Indians in the southwestern USA, Rift Valley fever among humans in Kenya in the late 1990s, and encephalitis deaths in New York City in September 1999 have converted the science fiction aspect of infectious diseases epidemics into non-fiction.

Political and scientific concern about the possibility of a global warming of the Earth's atmosphere has also sparked interest in societal vulnerability to infectious diseases. Whether global warming proves to be human-engendered or natural in origin is of no concern either to the pathogens (microorganisms) or the vectors (carriers) of those diseases. If the climate conditions change at regional and local levels because of global warming, various infectious diseases would adjust (shifting or spreading) to the newly developed hospitable habitats. Today, potential change in the interactions among climate, disease, and public health is one of the most dangerous, worrisome, and scary consequences of global warming.

Many of the infectious diseases that societies are worried about, if the climate in North America or in other industrialized countries were to change, already exist in abundance, often unchecked, elsewhere around the globe. They have neither been eradicated nor contained, despite major health campaigns to do so, especially in tropical countries (see Figure 12.1). They remain serious health threats in regions where the disease vectors are endemic; that is, where they are well suited to the local environmental conditions. In fact, the incidence of several diseases (malaria, yellow fever, cholera, dengue fever, tuberculosis) has been on the increase in recent decades.

Climate change means change in the various factors that combine to make up a region's meteorological conditions, e.g., precipitation, temperature, humidity, wind speed and direction, weather extremes. Scientists believe that global warming will make the hydrological cycle more active, producing about 15% more precipitation on a global basis. Some areas would become drier, and others wetter. Such changes in precipitation are expected to have a significant impact on the location, frequency, intensity, and duration of infectious disease outbreaks, because they will create new habitats for pathogens and vectors. No country, city, group, or individual will be immune to the public health-related changes that would accompany global warming, although populations will certainly differ in their capacity to adapt. Where these hydrological changes might occur is still subject to educated guessing.

Climate variability from season to season and year to year is known to

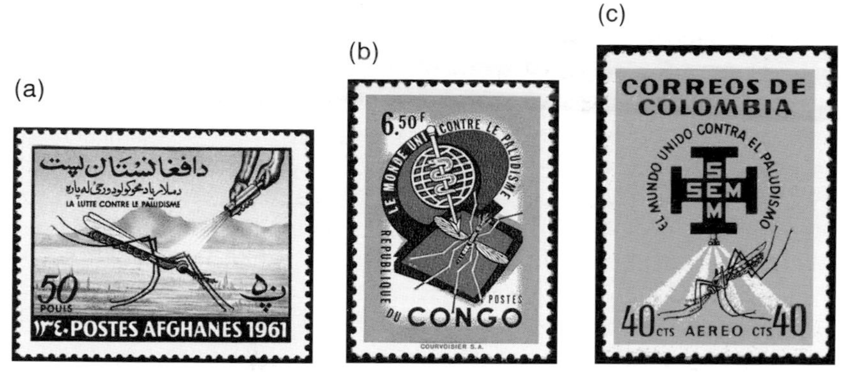


*Figure 12.1. 1961 stamps from the global campaign to combat malaria: (a) Afghanistan, (b) Republic of Congo, and (c) Colombia.*

affect human health by (a) its impacts on food production and water supplies or (b) its effects on the risk of diseases associated with various pathogens or vectors such as mosquitoes, ticks, and rodents. The following sections focus on the latter.

*Climate affects health in specific locations* Although infectious diseases are endemic to specific environmental conditions, with pathogens and vectors dependent on seasonal conditions for their survival, they can thrive under a wide range of conditions – from microenvironments (open containers such as cans or bottles, or puddles of standing water) to large areas. Thus governments and individuals have come to accept their coexistence with local disease pathogens or vectors, given their great difficulties in combating infectious diseases. That combat requires the constant monitoring of changes in the environment (especially the climate) and in the pathogens or vectors. Even though preventive measures and cures may exist elsewhere in the world, the costs to developing countries for these cures, or to individuals within them, are often prohibitive. Making a difficult situation even more so, microorganisms as well as vectors (mosquitoes in particular) often develop immunities to the cures. As a result, new strains are constantly emerging for which new cures are needed. For many people in the tropics, there is simply no escape from exposure.

*Movement of people into areas with vectors* Today, there is considerable movement of population into and out of areas with known infectious diseases. People move from one location where they already know what most of the health hazards are, to another location where they are likely to be unaware of the health problems.

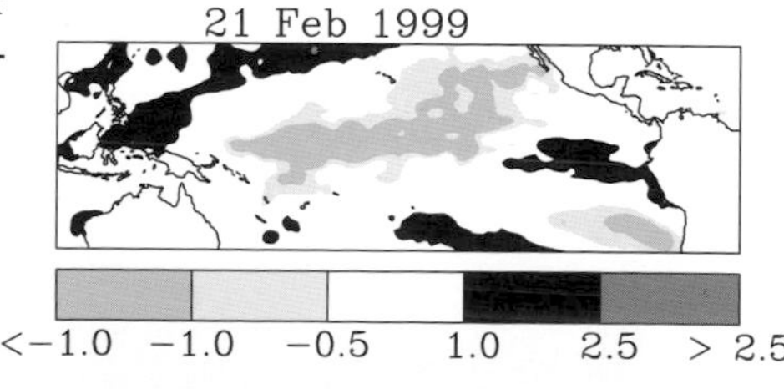

Some people move voluntarily while others are forced to do so, as a result of political, ethnic or military conflict, into areas that put them at high risk for diseases. Today many of the millions of people displaced by war have sought food and shelter in refugee camps. In these camps, overcrowding, poor sanitation, and poor nutrition combine to set the stage for infectious disease outbreaks. The resulting outbreaks capture the attention of the media, as was the case in Ethiopia's Korem refugee camp in 1984. In that instance, the international community had heard about famine within camps filled with starving peasants but failed to act on it, until a BBC television crew managed to take videos of the victims and smuggled the video to England. The showing of the video sparked public reactions that pressured Western governments politically opposed to the survival of the Marxist Mengistu regime to provide it with humanitarian assistance.

There are also many examples of governments forcing large groups of people into new settlements, which have often been in areas where vector-borne diseases thrive. Again, for example, the Ethiopian government of Col. Mengistu forced the large-scale resettlement of people from the country's highlands to the disease-plagued wet lowland areas bordering the Sudan. As expected and projected, a large portion of the newly resettled population fell victim to various water- and vector-borne diseases, and with the fall of the Mengistu government, many of the forced settlers returned to their home villages in the highlands.

*Vectors move into new areas inhabited by people* With global climate change, higher temperatures, more (or less) precipitation and more (or less) humidity will alter the risk of infectious disease outbreaks by, for example, changing "the potential length of the malaria transmission season, vector species distribution, and the suitability of particular malaria control options in that area" (Conner *et al.*, 1998). The geographic expansion of these infectious diseases will increase the risk of epidemics to areas where it had not previously existed. In addition, people in these infested areas lack the natural immunity to the newly introduced infectious disease.

As noted earlier, the recent outbreak in New York City of a certain strain of encephalitis never before seen in North America has been blamed on the increase in international air travel and the increased risk of transporting vectors (mosquitoes) on board flights from Africa, in this instance.

## *Contemporary efforts to understand climate and health*

A series of United Nations (UN) conferences was convened in the 1970s, sparked by international concern about population growth rates and global food shortages, as a result of drought. The first World Climate Conference was next to the last major UN conference of that decade. While the climate and health issue was one of the topics addressed in that conference, the concern at the time was with the ability of humans to cope physiologically with the effects of weather, heat waves, and air pollution – a subfield of research called biometeorology. Only a brief mention was made of climate and infectious disease. However, at the least, it was one of the few climate-related issues chosen by governments for discussion at the conference. With the benefit of hindsight, one can see that the health presentation was Eurocentric as well as narrowly focused.

A decade later, the Second World Climate Conference was convened in Geneva (WMO, 1991). Once again, climate and health issues were discussed. This time, however, global warming was considered along with biometeorological issues. Interest in climate change had grown sharply in the 1980s and was a key concern to participants at the Conference. While specific diseases were mentioned in a climate and health paper, no reference was made to El Niño's impacts on health, despite the existence of a few research articles on its health effects in 1982–83 (Caviedes 1985; Gueri *et al.*, 1986; Telleria, 1986).

In the 1990s, interest expanded rapidly with regard to climate and health interactions. UN agencies joined together to produced a state-of-knowledge report on climate–health in response to growing international concern about climate change impacts on human health (McMichael *et al.*, 1996). The report addressed weather and climate factors that affect human health (temperature, precipitation, humidity, air pollution), vector-, water- and air-borne infectious diseases (malaria, yellow fever, dengue fever, cholera, etc), food production and nutrition, the health impacts of extreme weather events, and climate change's possible health effects. The report also highlighted some of the impacts of recurrent El Niño events on human health. The box on p. 182 is reproduced from the report.

This multi-agency review drew attention to the "indirect effects upon human health arising from climatic stresses upon the stability and productivity of ecological systems ... and to the longer-term implications for human health of disturbing or damaging components of the biosphere" (McMichael *et al.*, 1996, p. 297).

## Infectious diseases

A recent workshop report on the effects of climate and the incidence and distribution of infectious disease discussed various microbiological changes that might accompany global warming (Colwell and Patz, 1998). The report suggested that

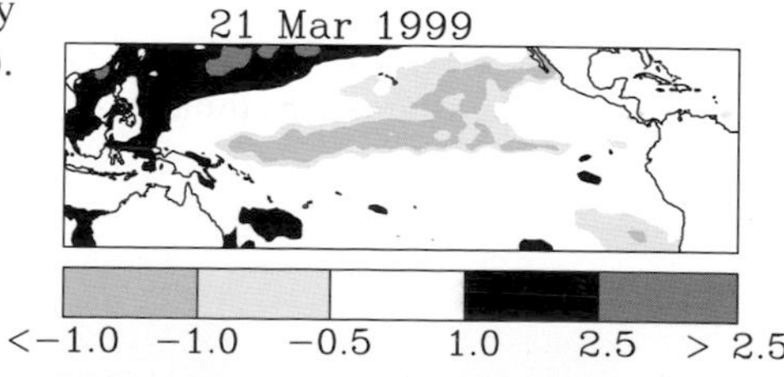

The following examples involving ENSO events illustrate that, for some diseases, changes in climate variability associated with climate change may be more important than changes in mean climate. Incidence of vector-borne diseases is often associated with the occurrence of climatic extremes such as excessive precipitation. In temperate southeast Australia, infrequent but severe epidemics of Murray Valley encephalitis (which can cause serious brain damage) occur after extended precipitation and flooding; heavy precipitation causes a rapid increase in the population of the mosquito vector. Many outbreaks correlate with ENSO phenomena. The Southern Oscillation Index, one of the ENSO parameters, which relates to air pressure deviation, can be used to predict the probability of an epidemic. Other vector-borne diseases for which a strong association between outbreaks and El Niño years has been observed include: eastern equine encephalitis in northeast USA, epidemic polyarthritis caused by the Ross River virus in Australia, and malaria.

Many areas across the globe that are affected by malaria experience drought or excessive rain coincident with ENSO teleconnections. Quantitative leaps in malaria incidence in Costa Rica and Pakistan, for instance, are coincident with ENSO events. Historically, in the Punjab region of northeast Pakistan the risk of a malaria epidemic increases five-fold during the year following an El Niño year, and in Sri Lanka the risk of a malaria epidemic increases four-fold during an El Niño year. These increased risks are associated with above-average levels of precipitation in the Punjab and below-average levels of precipitation in Sri Lanka (primarily because of the differences in mosquito species' ecology).

McMichael *et al.* (1996)

> Infectious diseases that are responsive to climate can be divided into two groups. The first group comprises those diseases for which there are clearly documented links between incidence and climate and weather factors. This group is primarily composed of vector-borne diseases, including malaria, hantavirus, pulmonary syndrome, dengue, and various forms of viral encephalitis . . . The second group comprises diseases whose incidence is cyclical (or seasonal), thereby suggesting a link to climate, but for which the potential mechanisms are either unknown or only tentatively established (such as influenza).
>
> (Colwell and Patz, 1998, p. 7)

Cholera, the report noted, is a disease in the second group. It is one of the most widespread of the water-borne diseases and its incidence was on the rise at the end of the twentieth century. The spread of this infectious disease is a major concern to those investigating how public health might be adversely affected by global warming. Cholera outbreaks in some countries have also been associated with El Niño events.

The diseases in the first group are vector-borne where microorganisms find hosts that provide optimal environments for their development. Those hosts also serve as a transport mechanism by which the disease is carried from one location to another. The two most frequently mentioned vectors are mosquitoes and rodents. Mosquitoes are vectors for malaria, dengue fever, and yellow fever. Malaria, which is endemic in the tropics, is one of the most climate-sensitive vector-borne diseases. In a report on malaria, the American Association for the Advancement of Science (AAAS, 1991, p. ix) noted that "malaria is a complex problem for which there is no magic bullet, no quick or easy solution, particularly in Africa, where approximately 80 to 85 percent of cases and 90 percent of deaths in the world due to malaria occur". As for rodents, they carry plague-bearing fleas and are linked to hantavirus outbreaks. According to the World Health Organization (McMichael *et al.*, 1996, p. 76), "Diseases involving vectors such as insects and rodents respond quickly to climatic, ecological, and social change".

## El Niño and health

Perhaps the first papers that focused specifically on El Niño's impacts on human health appeared in the journal *Disasters* in October 1986. Various articles in this issue focused on water- and vector-borne disease outbreaks in a few South American countries during the 1982–83 El Niño. Until the mid-1980s, most countries had paid little attention to the phenomenon or its societal impacts, let alone its health effects – even those countries known to be directly and adversely affected by El Niño.

Interest in the relationship between El Niño and health increased sharply during the El Niño event(s) of the 1990s (the long El Niño event(s) of 1991–95, and the very strong 1997–98 El Niño). Since the early 1990s, El Niño events have been used to provide insights into how well societies respond to interannual variability and examples of what might happen locally as a result of global warming.

The 1991 cholera outbreak in Peru and the ensuing cholera pandemic in South America (Epstein *et al.*, 1995), cholera in Bangladesh (Colwell, 1996), and the apparent association of El Niño-linked increases in dengue fever cases in Vietnam (Lien and Ninh, 1996) provide examples of how infectious diseases might spread if the global climate were to heat up. Despite the perceived linkages between climate variability, climate change, and El Niño, it is generally recognized in the climate–health community that research on those linkages between ENSO extremes and human disease outbreaks is in its infancy (e.g., Kovats *et al.*, 1999).

During El Niño, diseases in some locations tend to spread into new areas, at least temporarily. They can also become more

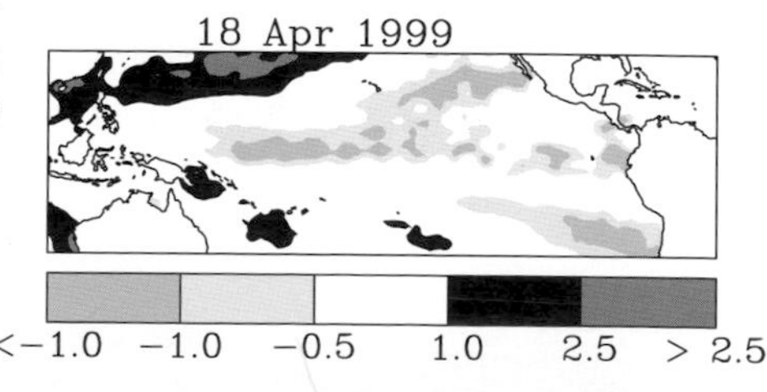

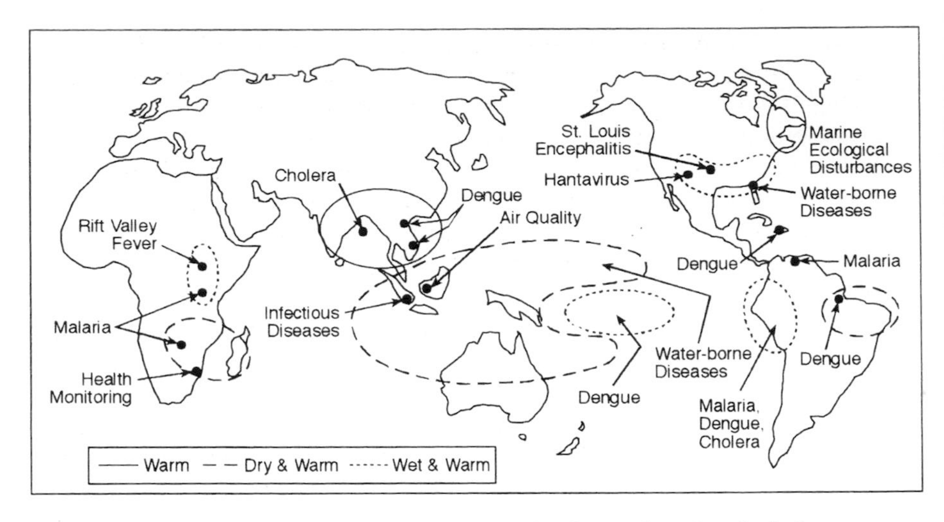


*Figure 12.2. The ENSO Experiment was designed to identify the key health impacts of the 1997–98 El Niño event. (Adapted from NOAA/OGP website (www.ogp.noaa.gov/mpe/csi/appdev/health/ensoexp.html).)*

prevalent in endemic areas, producing epidemics. Societal preparation for, or responses and adjustments to, such drastic impacts today can be instructive on how vulnerable (or resilient) societies might be in the face of health impacts of global climate change. Regardless of whether the climate changes or not, governments and individuals will still have to cope with infectious diseases from a variety of causes including, but not limited to, climate variability.

Epstein and his colleagues, who have studied disease and El Niño, noted that "changes in temperature, precipitation, humidity and storm patterns often related to the El Niño–Southern Oscillation (ENSO) phenomenon, are associated with upsurges of water-borne diseases such as hepatitis, shigella dysentery, typhoid and cholera" (Epstein *et al.*, 1998, p. 1737). As a result of Epstein's medical assessments of climate's impacts on health, he has suggested that El Niño forecasts can serve as a "weapon" against infectious disease. Figure 12.2 depicts the regions affected by the various infectious diseases under the 1997–98 El Niño.

### *Malaria*

Many countries are at increased risk for malaria during El Niño, e.g., Venezuela, Colombia, Pakistan, Ecuador, Bolivia, Sri Lanka, India,

Ethiopia, East Africa. A report from Medécins sans Frontières (Doctors Without Borders) offered some insights into what one might expect in El Niño years with respect to the incidence of malaria (Kovats *et al.*, 1999). For example, Kovats and colleagues noted that during El Niño years, higher temperatures, particularly during autumn and winter months, are likely to increase the transmission of malaria in the high altitude and high latitude areas of Asia. As another example, the minimum temperature for development of malarial parasites *Plasmodium falciparum* and *P. vivax* approximates to 18 °C and 15 °C, respectively. In the typically non-malaria endemic highlands of Kenya (Garnham, 1948), Rwanda (Loevinsohn, 1994), and Zimbabwe (Freeman and Bradley, 1996), increases in ambient temperature have been linked to epidemics of malaria. Also, the incidence and prevalence of malaria is closely associated with altitude, which is a good proxy measurement for temperature (Taylor and Mutambu, 1986).

Much of the increased attention to El Niño's impacts on health has focused on diseases associated with mosquitoes (malaria, dengue fever, and yellow fever) and rodents (carriers of plague-bearing fleas). When temperature and precipitation increase greatly from normal conditions, the likelihood of epidemics increases, at least at the local level and especially at the fringes of regions within which these pathogens or vectors are at home (i.e., endemic). Often the human populations along the fringes, where such diseases may be rare under normal conditions, are adversely affected as changes in temperature or moisture associated with El Niño events enable them to move into adjacent areas.

Temperature also has a direct impact on the various stages in the life cycle of mosquitoes, as well as on disease vectors and the microbial pathogens they may carry. For example, higher temperatures within a range in which mosquitoes reproduce can accelerate the timing and rate at which they breed.

Malaria is affected by decreases as well as increases in normal rainfall. A reduction in rainfall in a normally wet area may cause rivers to dry up, leaving pools of standing water that provide ideal breeding grounds for mosquitoes. In normally dry areas, anomalously heavy rainfall events can also create pools of standing water. These too can serve as breeding grounds for vectors and, as a result, vector-borne diseases (e.g., Bouma and Dye, 1997). For example, Telleria (1986) identified in Bolivia an increased incidence of malaria, along with other diseases, as a result of the 1982–83 El Niño.

Bouma and colleagues (1997) assessed the incidence of malaria in various parts of Colombia related to El Niño. They noted that two-thirds of the Colombian population live in areas that are endemic for malaria. The geography of the country is very diverse, making

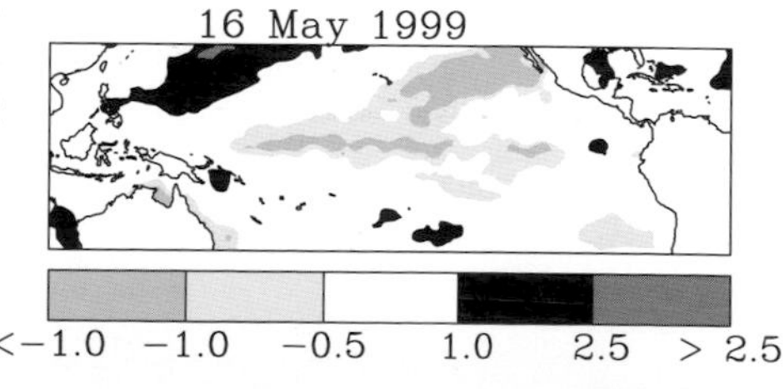

some locations more vulnerable than others at different times of the year. Also, in recent decades, there has been an increase in malaria in part because of the country's deteriorating public health facilities. They found a strong statistical correlation between El Niño and malaria, showing a 37% increase in malaria cases in the year following an El Niño (called El Niño year (+1)) as opposed to other years of normal and La Niña-related weather in the country.

Considerable interest has developed in using satellites to monitor environmental changes that increase the risk of malaria. By identifying and mapping existing vegetation cover (grasslands, forests, wetlands, etc.), rainfall and temperature conditions in a given region, and departures from average conditions (and from societally accepted levels of infectious diseases), such information can be used as an early warning of an increased risk of infectious diseases.

Satellite images can be used to assess disease risk over remote regions where meteorological data may be either unavailable or uncollected. In such locations, remote sensing by satellite can determine soil moisture content (Washino and Wood, 1994); a vegetative index and land surface temperatures provided by satellite are used to estimate evapotranspiration. Daily temperature differences obtained from satellites have been used as a surrogate for soil moisture to predict the prevalence of bancroftian filariasis caused by mosquitoes carrying the larvae for the filarial worm *Wuchereria bancrofti* in the Nile delta (Thompson *et al.*, 1996). In sum, remote sensing is already being used to predict malaria transmission in several endemic regions (Beck *et al.*, 1997).

### *Dengue fever*

Dengue fever outbreaks are a public health risk in wet tropical regions, especially those areas affected by high rainfall during either an El Niño or a La Niña. Kovats and colleagues (1999, p. 12) noted that "increased rainfall in many locations can affect the vector density and transmission potential". While dengue is a problem during El Niño in South and Southeast Asia, it is a problem in the Caribbean during La Niña.

Like those that carry malaria parasites, the mosquitoes that carry the dengue fever virus are also sensitive to temperature. While freezing temperatures kill larvae, higher temperatures increase and enhance the vitality of the dengue fever pathogen by affecting its replication, maturation and period of infectivity. The warmer the temperature up to a point, the faster they develop and the earlier in their life cycle they can infect (Colwell and Patz, 1998).

An estimated 2.5 billion people, mainly in the tropics and the subtropics, are at risk for dengue fever, and there are an estimated 250 000 to 500 000 cases annually. Taken globally, "dengue viruses are one of the most important arthropod-borne viruses transmitted to humans" (Patz *et al.*, 1998, p. 147). There is no vaccine for dengue. As bad a threat to human health as dengue fever is today, climate change scenarios suggest a much more ominous picture. A recent study linking potential transmission risks to general circulation models of global climate change suggested that a global warming of a few degrees Celsius "may allow dengue and other climate sensitive vector-borne diseases to extend into regions previously free of disease, or they may exacerbate transmission in endemic parts of the world" and (Patz *et al.*, 1998, p. 147).

### *Yellow fever*

Yellow fever is yet another vector-borne disease with some of its outbreaks linked to El Niño. Once considered to be on the road to eradication by the World Health Organization, yellow fever appears to be on the increase once again in South America and in parts of Africa. Researchers have suggested that the risk of yellow fever in West Africa doubles during El Niño events (Kovats *et al.*, 1999). Although a yellow fever vaccine is effective, it is not available throughout the world when yellow fever outbreaks arise as a problem.

## Concluding comments

The forecasting of the effects of El Niño and La Niña on temperature and rainfall in regions prone to the outbreak of various diseases, whether vector- or water-borne, can provide enough early warning for the preparation of measures to reduce exposure to those diseases. However, it will take a major effort on the part of the rich countries ("the haves") to provide the necessary vaccines and to assist with other preventive measures in developing countries ("the have nots") in the face of El Niño or La Niña forecasts. There are many environments (from micro to macro) in which vector-borne pathogens could (and have) survive(d), if certain conditions prevail. A few examples hopefully will provide a glimpse of how El Niño might (or has) affect(ed) the incidence of different infectious diseases.

It is important to note that there may be disadvantages related to the overreliance on the use of El Niño forecasts for the health sector such as the following: there is limited reliability of El Niño forecasts; there are varying intensities and therefore impacts of El Niño; there are stronger El Niño teleconnections in some locations than in others; there is a lack of reliable health data over long periods of time; the causes

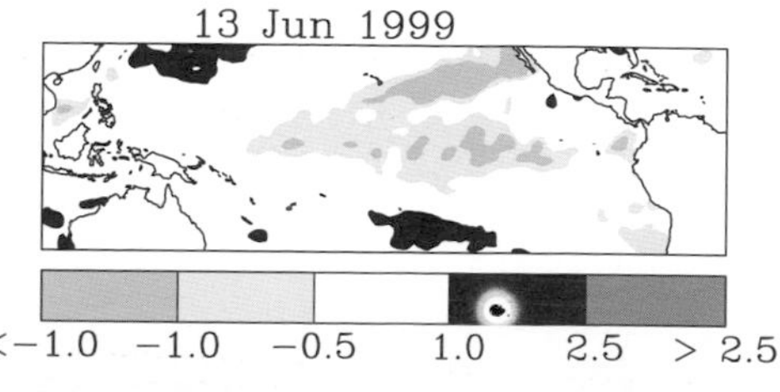

of epidemics and disasters are multifaceted; health data are difficult to interpret (Nicholls, 1988).

Kovats and her colleagues (1999, p. 8) suggested that for studies linking El Niño to human health impacts to be convincing, "they must be able to satisfy the following criteria: (1) there must be climatological evidence of robust teleconnections to the region of interest with El Niño in the Pacific, (2) there must be biological evidence that the diseases or other health impacts of interest have a plausible biological link with weather exposures, (3) there must be epidemiological evidence (i.e., statistical analyses that show that the incidence of specific diseases in specific locations vary together over time)".

With regard to these and other diseases, societies are not blameless. Mosquitoes are opportunistic, finding breeding grounds wherever they can. For example, open containers that capture water serve as excellent breeding grounds. Health services in many countries are of poor quality, given the lack of foreign currency to purchase appropriate prophylactics to prevent, if not cope with, malaria epidemics that might occur in various microenvironments. It is interesting to note that, at the turn of the nineteenth century and into the early part of the twentieth, both malaria and yellow fever were endemic to the USA but were then eradicated. Thus societal intervention can help to minimize if not eliminate the existence of these disease vectors. On the other hand, public health expert Jonathan Patz (personal communication, February 1999) noted that, "complacency in vector borne disease control and surveillance has partly led to a resurgence of many diseases, and some vector species, i.e., [*Ades*] *aegypti*, have proved nearly impossible to eradicate".

# 13 The media, El Niño, and La Niña: a study in media-rology

> Media coverage of the 1997–98 tropical ocean warming event made the term "El Niño" a household word. So pervasive was coverage of El Niño that it became the fodder of late-night talk-show monologues and an oft-invoked gremlin responsible for many of society's ailments.
>
> (Hare, 1998, p. 481)

## Introduction

Broadcast and print media are important conduits that pass scientific information from researchers to the public. They do so for altruistic reasons as well as for self-interest. In the process, they often have to translate scientific jargon into popular language and to present it within a short space or allotted time. This process is good for educating the public on scientific issues as well as for providing early warning to decisionmakers of the possible impacts of the phenomenon about which they are writing. The late world-renowned American scientist and author Carl Sagan, in a foreword to a guide book on science writing, suggested that "The writers of articles and books on science (and their radio and television counterparts) are the chief means by which adults in our society learn about science" (Sagan, 1997, p. vii). Another science writer (Rensberger, 1997, p. 8) observed that "Newspapers . . . are the front lines of science communication, the place where most science stories show up first, before they appear in magazines, long before they're in books, usually years before television documentaries discover them."

In addition to being educators of the public, the media are businesses that compete with others in the same line of work for attention and for a share of the consumer market and revenue. The more people they can persuade to use their services, the more likely it is that their profits will increase. That is the way the media operate in a democratic society. That is the way it is when it comes to writing about scientific issues such as those related to El Niño and La Niña.

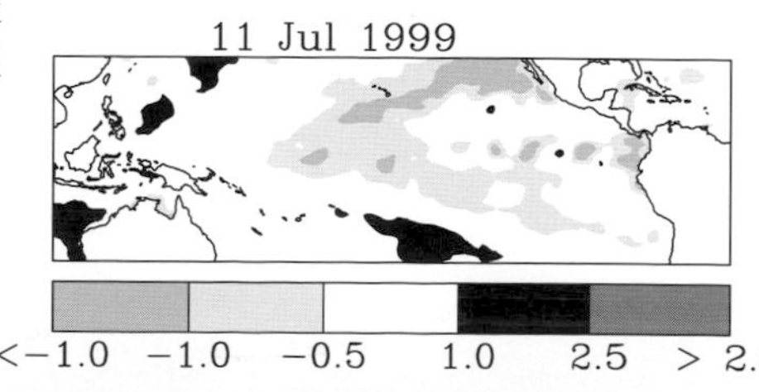

Thus it is advantageous for the media's science writers to present scientific programs, issues, and research in a timely way for both business and educational purposes. Such writings are also of great value to the scientists engaged in the research activities that the media choose to highlight. On this point, a science writer suggested that

> Newspaper science writers are what the academics call gatekeepers, people whose jobs allow them to decide what developments in the real world get into the news, and hence reach the public. The public relations types at every scientific and medical organization know that if they can win [media] attention and interest, they will have scored. And they know that if the newspapers do their story, the other media are likely to follow.
> (Rensberger 1997, p. 7)

While there may be complaints from various groups within the El Niño-related scientific community (i.e., researchers and forecasters) that the media stories have not accurately, adequately, or properly covered the El Niño phenomenon, there are many other fields of research that would love to get the high level of attention and publicity that El Niño research and researchers were receiving during the 1997–98 event, the last and biggest El Niño of the twentieth century.

## El Niño and the media

Searching old newspapers and magazines for a mention of El Niño in the USA is an interesting exercise. Going back a few decades, one can hardly find a reference to El Niño. In those days, few reporters, if any, knew about El Niño, let alone how it might affect climate outside the tropics. Articles mentioning El Niño began to appear in the 1970s with greater frequency as a result of El Niño's adverse impacts on the Peruvian fisheries in 1972–73, and later in the decade because of increasing research activities. Media coverage in the 1970s, however, was still quite low and sporadic.

The media showed increasing interest with the 1982–83 event. Since then, media only sporadically "feasted" on El Niño stories sparked by the onset of an event. Media interest was piqued once again in the early 1990s. The 5-year period from 1991 to 1995 received considerable scientific attention because the Pacific sea surface temperature changes that began in 1991 did not follow the usual pattern. Although a typical normal El Niño took place in 1991–92, sea surface temperatures did not drop below normal for very long. They fluctuated above and below average again and again until early 1995. The anomalous behavior of the climate system was undoubtedly being affected by the large-scale volcanic eruption of Mount Pinatubo in the Philippines during 1991. The Pinatubo eruption put large amounts of particulates in the high levels of the atmosphere, which caused a cooling in the tropics and elsewhere for a few years.

Local media in the USA have shown varying degrees of interest in El Niño. In large measure that interest depended on whether local weather conditions might be affected by it. Media interest was clearly reinforced by numerous anecdotal and scientific comments on the local impacts, which affected Alaska, down the west coast of North America, across northern Mexico and the Gulf regions and the southeast, and up the east coast of the continent into northeastern Canada. Many of the anecdotes and comments received new levels of credibility as a result of the North American impacts of the 1997–98 El Niño.

Particularly since the 1997–98 event, the media in other parts of the globe have also given El Niño their fullest attention. The Brazilian media, for example, are concerned because parts of Brazil tend to be plagued by severe droughts (the poverty-stricken semi-arid northeast) and, at the same time, parts of the south are affected by conditions of favorable rains to heavy floods. Southern and northeast African media have also become aware of the increase in probability (but not certainty) of the devastation that can accompany El Niño. The media in Southeast Asian nations, such as Vietnam, Thailand, Indonesia, and the Philippines, pay close attention now to El Niño and La Niña forecasts as well.

The media also serve as educators of the public. Their articles and broadcasts generate interest and understanding about the phenomenon. Therefore it is crucial that the media have a correct understanding of what El Niño is, what it does, and what value there is to knowing about it. It is also imperative that the media improve their understanding of how such information might be of use to different sectors of society.

## Media interest in the 1997–98 El Niño

El Niño researchers and forecasters, relying on considerable amounts of information from the network of satellites, buoys, ships of opportunity, and computer and statistical models, came to realize as early as February–March 1997 that an El Niño was forming in the central Pacific along the equator. In March, an ENSO advisory was issued in the USA by NOAA (1997a), stating that the formation of an El Niño was in its early stages. In June, observations of sharply increasing ocean temperatures in the Pacific along the equator, along with several El Niño forecasts based on the outputs from computer models, prompted a second official news release by NOAA (1997b). This news release forecast the development of a strong El Niño later in the year, peaking in the winter of 1997–98. Claims to the media by scientists compared the emerging El Niño to the devastating 1982–83 event. As a result, major adverse impacts were

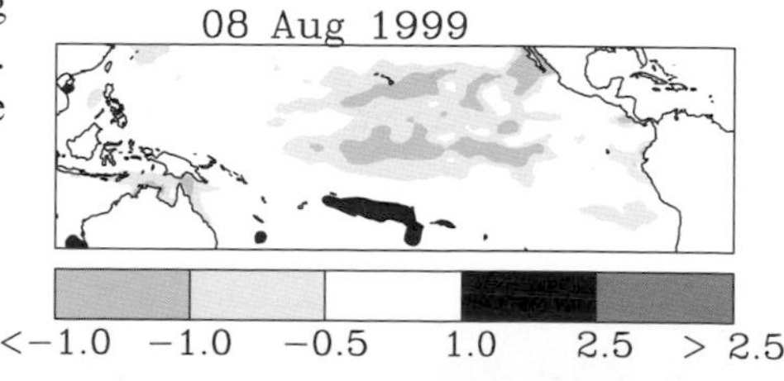

expected to accompany this El Niño in the Fall and winter of 1997 and into the spring of 1998.

In a review of media coverage of the 1997–98 El Niño on the Internet, Steven Hare (1998) identified a progression of changes in interest over time in the phenomenon. As a result of his review, he noted the following:

> The subject matter of the articles was highly varied and changed steadily as the event progressed. Early stories focused on predictions that the 1997–98 event would likely challenge the 1982–83 event as the strongest in modern history. By October 1997, stories started appearing on preparations being undertaken to minimize the impact of El Niño-fueled storms along with estimates of the potential worldwide damage in dollars. In November and December, stories about the actual impacts – speculatively linked to El Niño – dominated the stories. Impacts ranged from the fires in Indonesia to floods in South America to impacts on Pacific Ocean biota such as fish and birds. There was somewhat of a lull in January 1998, with a number of stories suggesting that the El Niño event was waning. In February, however, the number of stories skyrocketed as the Southwest and Southeast US both experienced record amounts of rainfall with widespread flooding and damage. In March, reports started appearing on the financial damage resulting from El Niño storms, along with stories about the accuracy of El Niño forecasts several months before the onset of the event. Reports in May about the decline in size of the warm pool were quickly followed by June stories about an impending La Niña event. The number of Internet stories increased again in July 1998 as more media coverage was devoted to La Niña. Much of the interest in La Niña was spurred by the NCAR-sponsored La Niña Summit held 15–17 July 1998.
> (Hare, 1998, p. 481)

It is now evident that the media played a major role in fostering awareness among policymakers and the public some months in advance of the onset of the 1997–98 El Niño and its potential for generating devastating impacts globally. As it happened, that event did turn out to be the strongest one in the twentieth century, surpassing the previous record holder: that of 1982–83. Reporting on El Niño became so intense and often so sensationalized that the media were accused of "hyping" El Niño in a negative way. The media for the most part generated "doom and gloom" scenarios, several of which were either premature or eventually proven to be incorrect. One media person suggested that "El Niño might as well have been the Spanish word for hype".

An Internet website reporting on the media wrote the following: "El Niño is a mass of unusually warm ocean water, causing upheaval in weather patterns. El Bunko is the mass media's usual hot air, causing perceptual distortions. While the damage done by El Niño can be estimated, the harm that results from the El Bunko effect is so extensive as to be incalculable" (Solomon, 1998). Even the media have assessed their performance during the 1997–98 El Niño. For example, an Associated Press story in early February 1998, *before* the heavy rains began to cause major damage in Southern California, noted that "the predictions came true elsewhere in the world from a tame Atlantic hurricane season to heavy rain in Central Africa. But sunny Southern California has lived through the early hype relatively rain free, raising the question: Is El Niño a bust?" (Allen, 1998, p. A6). Reporters, in defense of their coverage, have contended that the constant headline coverage of El Niño in the media had been dictated to them by the high level of public interest in the phenomenon.

A CBS News poll conducted in late November 1997 asked Americans whether "El Niño was the real thing or just hype". Its report noted that "while some skeptics say El Niño is little more than a media event, almost two-thirds of those aware of El Niño believe it will affect the weather where they live" (CBS News, 1997).

The Environmental News Network (ENN) conducted a survey among its staff members to identify the top 10 news stories of 1997. They voted El Niño as the number 1 environmental news story of the year, receiving just enough votes to surpass the global warming issue in general and the Kyoto Conference of Parties to discuss the protocol for the UN Framework Convention on Climate Change held in December 1997.

## Media hype: good and bad

The amount of American media interest in El Niño had grown sharply by late 1997. From a few El Niño stories early in the year, the phenomenon and its potential impacts were being discussed in nearly every newspaper and on every television and radio station by the year's end. The media (along with the public and decisionmakers) had become increasingly attracted to El Niño stories, especially when the emerging 1997–98 event was compared in magnitude and impacts to the devastating 1982–83 event. It was as if the media went into a frenzy over El Niño coverage. Headlines became increasingly provocative, often to the extent that they failed to reflect the story that they headlined (according to reporters, copyeditors usually wrote the headlines for their stories and resorted to catchy headlines to attract readers) (Figure 13.1). Media coverage of El Niño was high in Australia as well. With respect to El Niño stories in Australia, Nicholls and Kestin (1998, p. 419) observed that "At times . . . inappropriate headlines sometimes distorted messages, even if the text of an article was accurate".

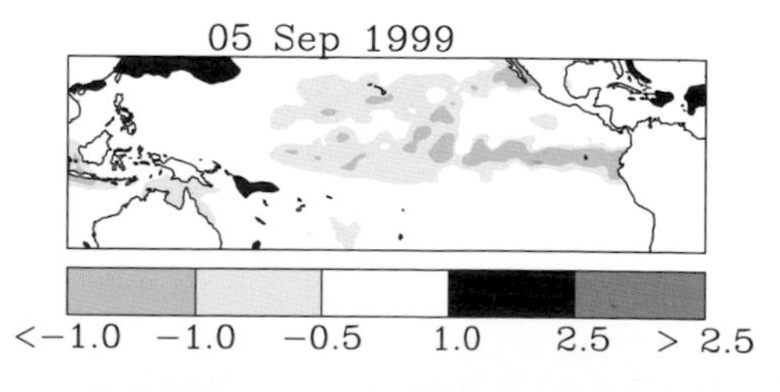

**hype** \ 'hip \ *vt* **hyped**; **hyp**'ing (1926) **1** : put on, deceive
**2** a: stimulate, enliven – usu. used with up b: increase gimmicks designed to ~ attendance at the games
**3**: to promote or publicize extravagantly – **hyped-up** \ ,hip-'dəp \ *adj*
*Merriam Webster's Collegiate Dictionary*, 10th edn, 1993

*Figure 13.1. (Reprinted with special permission of Cartoonists & Writers Syndicate.)*

*Figure 13.2. (Reprinted with special permission of* The Rocky Mountain News.*)*

*Good hype*

I consider "good hype" as referring to media coverage that educates the public about the El Niño phenomenon and generates awareness of its societal importance. It is not difficult to find media coverage that does this. Even those stories that do not capture the full details of the process or its impacts serve this purpose. The general public was learning by way of the media that El Niño is something in Nature that they should be concerned about, that it takes place in the Pacific Ocean, that there are fewer hurricanes in the Atlantic when an event is under way, that it returns every so often, that the Peruvians and the Australians are directly affected by it, and that the major fires in Indonesia in the Fall of 1997 were only partly related to El Niño. They also learned that El Niño was one of the biggest natural climate-related disrupters of human activities.

As a result of the 1997–98 event, numerous cartoons appeared for the first time

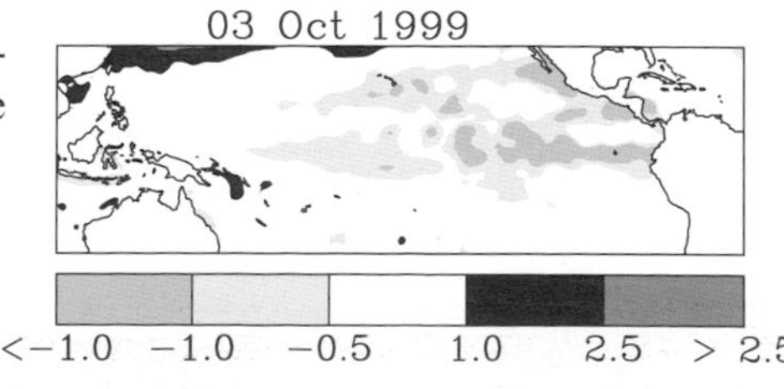

*Figure 13.3. Advertisements from the* Denver Post *newspaper, November 1997. (Reprinted with permission of* Rocky Mountain News.*)*

depicting El Niño (Figure 13.2). They proved to be another way to infiltrate the terms "El Niño" and "La Niña" into the consciousness of the general public. They fall into the good hype category.

References to El Niño also appeared in television commercials and newspaper advertisements. These, too, can be viewed as good hype. The advertisements in Figure 13.3 are typical of what appeared during the 1997–98 El Niño. Many of the suggested uses of the term "El Niño" were clearly in fun in this context and were not used by companies to "scare" consumers into buying their products.

*Bad hype*

I consider "bad hype" to be the use of sensational headlines to capture the attention of readers and viewers. It is often devoid of new, useful, or substantive information. For example, one of the major US television network news shows ran an *El Niño Watch* segment each night for many months. Toward the end of the El Niño event, the show's producers apparently searched far and wide to find a climate-related problem somewhere on the globe that they could blame on El Niño. In the early days of *El Niño Watch*, the show generated awareness, interest, and concern about the phenomenon and its potential regional and local impacts. By the end of the El Niño event, however, that nightly news show had turned much of that earlier interest into cynicism.

Scores of new El Niño-related websites began to appear. Many newspapers, radio, and television stations maintained El Niño updates on their websites along with animations and graphics. Some websites, however, chose to focus only on the media hype aspect of El Niño. One site, for example, went to the trouble of collecting various problems that someone, somewhere, at some time, had blamed on the 1997–98 event. Without explanation, this site listed the following "plagues" that had been blamed on El Niño:

> Acid rain, Bees, Black Scorpions, Black Widow Spiders, Bubonic Plague, Cholera, Crocodiles, Crop Failures, Dead Albatross, Dead Boobies, Dead Butterflies, Dead Coral, Dead Cows, Dead Dolphins, Dead Kelp, Dead Orangutans, Dead Penguins, Dead People, Dead Salmon, Dead Sea Lions, Dead Seals, Dead Terns, Dengue Fever, Drought, Encephalitis, Encephalitis (Equine), Encephalitis (Japanese), Famine, Fire, Floods, French Revolution, Frogs, Hantavirus, Heat, Hippopotami, Late Wine, Lyme Disease, Malaria, Meningitis, Mosquitoes, News Media, Rats, Rift Valley Fever, Smoke, Tornadoes, Unsold Pacific Tuna.

In sum, the media engaged in considerable competition for El Niño news. Each member sought a new angle for a unique story. The media had apparently become more enamored with its competition with other media over reporting on El Niño than with the value of the information they were providing. This, of course, is not unique to the world of science writing in North America. A cartoon that appeared in the *New Yorker* magazine captured the essence of what happens when the media fall into such a competition over news, even after the interest of the public in that news may have waned. The cartoon showed newscasters, each with his or her own microphone, in the face of the opposing television station's newscaster. They were backed up by television cameramen and

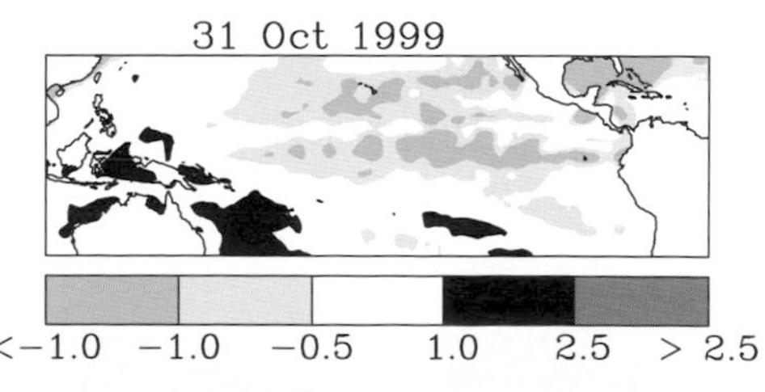

each station's reporter, taking notes. There was no caption, but the image was one of the depths to which newscasters had sunk in generating their own news by interviewing each other about a specific news event: no public, no politician, no expert, only newscasters interviewing newscasters.

## Where does El Niño science end and the reporting on El Niño science begin?

Thanks to the development of the TOGA-TAO Array and the Internet, scientists as well as the media, the public, and policymakers can monitor El Niño's and La Niña's growth and development day by day. Each wiggle and turn in the trend line of sea surface temperatures, however, now prompts a scientist somewhere to make a new projection to the media; the temperature trend line moves downward and she or he suggests that El Niño is weakening. Shortly thereafter it moves upward and she or he says that El Niño is strengthening! This is confusing not only to the general public and policymakers but to many researchers as well. It seems that once the media's interest has been piqued, there is a tendency for them to overfocus on the day-to-day changes in sea surface temperatures in one of the several sections into which scientists had divided the tropical Pacific Ocean (i.e., Niño1, Niño2, etc., see Figure 4.4).

Scientists are sought after by the media to comment on each twist and turn in the trend line of tropical Pacific sea surface temperatures or of sea level pressure changes (i.e., the Southern Oscillation Index). They (myself included) had been drawn into trying to explain each squiggle in a trend line for the 1997–98 El Niño, squiggles that can be observed but that may not be useful for long-lead forecasting. We had become describers of the daily routine of a natural process that is quite variable on different time scales. A good example of this occurred in November 1997, when NASA's Jet Propulsion Lab (JPL) issued a press release stating that the 1997–98 El Niño was starting to decay, based on a downturn in the trend line of some El Niño indicators. Two weeks later, JPL issued a new press release suggesting that El Niño was strengthening as the trend turned upward. Other research groups were baffled by such statements based on short-term shifts in El Niño's indicators. Recent comments by science writer Michael Lemonick (1997) on a different matter can shed some light by analogy on what might have been going on in this particular instance. Lemonick (1997, p. 202) suggested that "while it's very satisfying to come upon a piece of breaking news long before it breaks, it's even more satisfying to detect a trend well before it happens". Apparently this notion applies to El Niño scientists as well as to science writers.

In fact, many scientists have acted like El Niño reporters. Flattered to be asked by the media for their interpretations of each squiggle in a sea surface temperature trend line (i.e., interpretations of the week), some researchers have shown little reluctance to share their views of the moment with the media; first, about the importance of the phenomenon; second about forecasting El Niño's future development; and, third, about what its societal impacts might be. When pressed by the media for specific information about these impacts with regard to time and space (i.e., "What does El Niño mean to us in Iowa?"), researchers often speculate. Needless to say, their speculative comments make their way into the local news. As El Niño grows in size and strength, the adjectives that are used become more threatening.

As the expected consequences in the USA of the 1997–98 El Niño became more dire, the political aspects of El Niño emerged in media coverage: an official El Niño summit for the USA was convened in mid October 1997 in California. US Vice President Al Gore, US Federal Emergency Management Agency's (FEMA's) James Lee Witt, the state of California's Governor Pete Wilson and Senator Barbara Boxer, among others, got into the El Niño act for different reasons, mostly political.

By analyzing the ways the 1997–98 El Niño was discussed by scientists, policymakers, politicians, local weather forecasters, and the media, lessons can be found. For example, scientific researchers and the agencies that fund them must provide the media with a more realistic picture of what they know, as well as what they do not know but would like to know, about the El Niño phenomenon. For their part, the media have a responsibility to become better informed on the facts surrounding El Niño – about forecasting it, about the process by which it develops, and about its potential worldwide impacts.

Media representatives have repeatedly stated that a serious communication problem exists between them and the scientific community. Causes of this problem include, but are not limited to, the following: very few science writers have scientific backgrounds; there is a tendency among scientists and the media to overstate conditions (the media tend to highlight extremes and shifts, while the scientists tend to present their capabilities in a more positive light); a time constraint is an important factor because it is difficult for the media to maintain interest in long-term scientifically based issues. The media have the added difficulty of presenting a balanced range of views, having to take note of minority views on scientific issues, even when those views might lack much support. Referring to "the difficulties of communicating climate science through the media" to the public, Nicholls and Kestin (1998, p. 417) noted that "despite good will on all sides, the impression provided by media reports can be distinctly at odds with what scientists regard as the important messages". Using examples from public responses to the Australia Bureau of

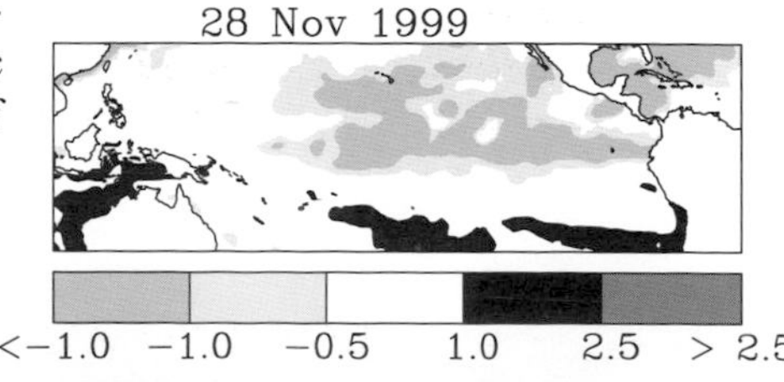

Meteorology's *Seasonal Climatic Outlook*, Nicholls and Kestin (1998, p. 418) also noted that

> Even ongoing contact [between scientists and the media] does not . . . ensure sound communication as another example from Australia in 1997 can illustrate. This time the scientists involved regularly wrote the media releases themselves. They were experienced in dealing with the media and had undertaken courses on dealing with the media. Despite this, the message that reached the community was distinctly different from the one intended by the scientists.

The following box provides highlights of an Australian case study of media interpretation and dissemination of the release by the Bureau of Meteorology of its seasonal outlook for the 1997–98 El Niño event.

From May 1997, the [Australian] Bureau [of Meteorology] had been including indications of a likely El Niño event, and hence of an increased probability of low rainfall over eastern Australia (the part of the country usually affected by the El Niño) in the media releases.

The Outlook issued in early August is probably representative of the Outlooks through the period May–November. Its headline was "El Niño persists: Dry weather likely to continue over southeastern Australia". The summary went on to say that "there is a strong likelihood of significantly drier than normal conditions persisting and expanding across much of eastern and southern Australia." . . . Rainfall was good through much of the region in September, a critical time for crops.

Towards the end of the year, criticism of the Bureau's Outlooks was reported in the media. The thrust of this criticism was that the Bureau should have been clearer in describing the limits to the accuracy of its predictions. . . . The use of words such as "likely," it was thought, indicated that the forecasts should not be interpreted as a categorical prediction of drought. However, the media criticism, and discussions with forecast users, indicated that there was a wide gap between what the Bureau was attempting to say (i.e., an increased likelihood of drier than normal) and the message received by users (i.e., definitely dry conditions, perhaps the worst drought in living memory) . . .

Part of the problem is attributable to the different emphases placed by forecasters and users on certain critical words. It appears that users and forecasters interpret "likely" in different ways. Those involved in preparing the forecasts and media releases intended this to indicate that dry conditions were more probable than wet conditions, but that there was still a finite chance that wet conditions would occur. Many users, it appears, interpreted "likely" as "almost certainly dry, and even if it wasn't dry then it would certainly not be very wet". This interpretation appears to have been reinforced by the plethora of media reports on the El Niño . . .

One problem with the media releases for the Outlooks is that they are, by necessity, short. Much of the detail included in the Outlooks (e.g., the tables of probabilities) cannot be included in the media release. So words such as "likely" are used to replace the calculated probabilities . . .

We point to a successful use of the Bureau's Seasonal Outlooks and ancillary information during 1997–98 in reducing the danger of bush fires in southeastern Australia. In the middle of 1997, after the Bureau had begun issuing forecasts regarding the possible impact of the El Niño, rural fire authorities contacted Bureau personnel for further information. This action was prompted by the massive damage and loss of life caused by bush fires during the 1982–83 El Niño event. An intensive media campaign was initiated to alert people to the possible danger from bush fires because of the dry and hot conditions that often accompany El Niño events. In the event, several days in February and March 1998 had conditions very conducive to bush fires, quite similar to the conditions in 1983, yet fire damage to residential property was limited (although considerable forecast areas were burnt). At least part of the reason for the relatively low level of damage was the publicity campaign that sprang from interactions between forecasters and users (rural fire authorities) . . .

It is crucial, therefore, that organizations and individuals with a climate message develop improved methods for delivering their message through the media, despite the pain.

Kestin and Nicholls (1998)

It is imperative that scientists and the media devise ways to interact that are effective for both communities, exploring new avenues for each to educate the other more effectively on the information related to the ENSO warm and cold event cycle.

The scientific community "discovered" the global implications of El Niño only as recently as the mid-1970s. Scientists still have much to learn about the science and impacts of this natural phenomenon. Given its importance, it is imperative that the scientific community do a better job of educating the media about El Niño so that when the next extreme warm or cold event does occur, most likely in the first couple of years of the twenty-first century, societies around the globe will be much better educated and, therefore, better able to cope with El Niño when it peaks. They will also be better prepared to use the earliest warning of adverse climate conditions that ENSO forecasts provide along with real-time observations and forecasts of its progress.

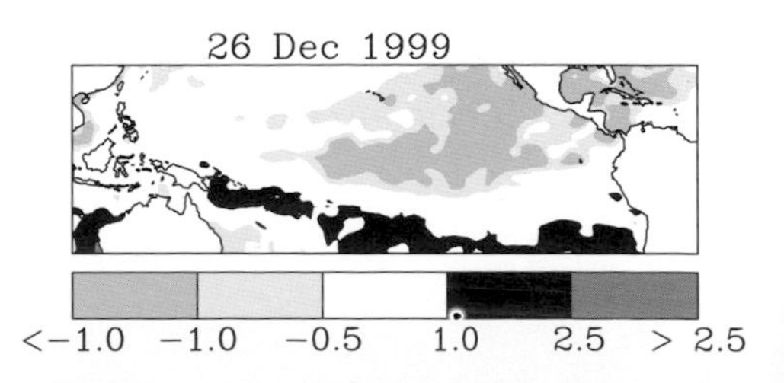

# 14 Why do ENSO events continue to surprise us?

I really thought that record would stand,
until it was broken

(Yogi Berra)

## Introduction to surprise

People are being surprised every day but not necessarily by the same things, at the same time or in the same way. With respect to El Niño, it seems that the experts – those whose scientific energies have been focused on understanding, explaining, and forecasting this phenomenon – have been surprised for one reason or another by each successive El Niño up to the present. In one sense, people know that an El Niño will return to the tropical Pacific. In a way, then it is a "knowable" surprise. What people do not know are what its characteristics will be when it returns – when it will return, how strong it will be, what impacts it will have and where, and which regions in a given country will be most affected.

Thus surprise often arises because people do not know what the changes in the various aspects of a climate-related event are likely to be (e.g., timing, intensity, duration, location, likelihood of occurrence, impacts). Surprise can also arise as a result of the way that people interpret or perceive the world around them; the way they process historical information, the way they deal with probabilities, their understanding of the climate system, and the way they view the risk of rare, record-setting weather events. By using any one or a combination of a variety of methods – scenario-playing, computer modeling, historical review, paleoclimatic reconstruction, geological assessment and forecasting by analogy, among others – some *potential* physical climate surprises (type A, physical) can be identified in advance as can some of the *potential* environmental and societal impacts (type B, societal) that result from those surprising changes in the climate.

In general, a climate surprise can be viewed as a gap between what one expects the climate to do and what the climate actually does. T. R. Stewart (1997, p. 7) wrote that

> surprise is clearly subjective because what is surprising to one person is not always surprising to another. The expectations that are necessary for surprise are also subjective. To say that they are subjective, however, does not mean that they are totally independent of reality. Expectations are formed in different ways, depending on whether we face a familiar situation or a new situation. In familiar situations that we encounter repeatedly, we acquire expectations through learning . . . We also have expectations in unfamiliar situations . . . If we do not find what we expect, we are surprised.

## Defining surprise

The *Oxford English Dictionary* (1971), which traces word-use centuries back in time to its origins, defines "surprise" in these various ways: surprise is "an act of assailing or attacking unexpectedly without warning"; "anything unexpected or astonishing"; "the feeling or emotion excited by something unexpected or for which one is unprepared"; "unexpected occurrence". As one can see from this sampling, there is a common element in each definition of surprise – the unexpected – but definitions of surprise also contain words such as "astonishing" or "unprepared". So, an event can be unexpected for a variety of reasons.

Ecologist C. S. Holling (1986, p. 294) noted that surprise involves both natural and social systems, suggesting that "surprises occur when causes turn out to be sharply different than was conceived, when behaviors are profoundly unexpected, and when action produces a result opposite to that intended". He emphasized the importance of expectations. Stewart (1997; see also Timmermans *et al.*, 1996) noted that

> since climate is changing, and the climate record is short (except for some tree-ring and ice-core studies), objective probabilities are rarely available. As a result, the public and policy makers must rely on their own judgment and apply their probability estimates to guide their climate-related decisions. Research has shown that the accuracy of probability estimates of both experts and lay people in general is often inadequate.
>
> (Stewart, 1997, p. 4)

Over the years, various researchers have related surprise to all kinds of change: global change, climate change, environmental change, economic change, cultural change, technological change, military tactics and strategies, and even humor. Harvey Brooks (1986, p. 326), for example, writing about surprises "in relation to the interaction

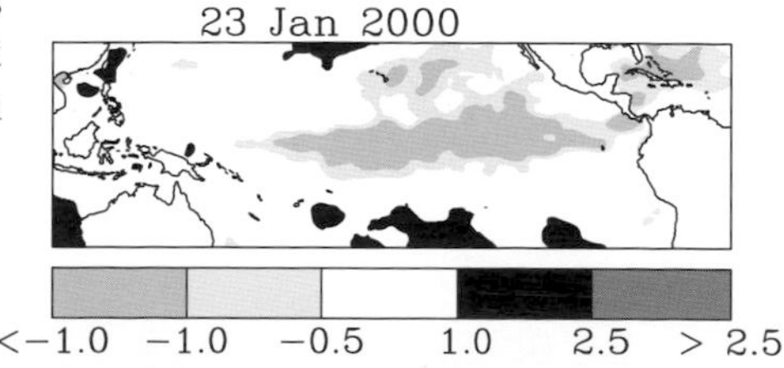

between technology, human institutions and social systems," clustered types of surprise into (a) unexpected discrete events, (b) discontinuity (sharp breaks) in long-term trends, and (c) the sudden emergence into political consciousness of new information.

Thompson and his colleagues (1990), on the subject of the cultural aspects of surprise, identified three important points: (a) an event is never surprising in itself, (b) an event is potentially surprising only in relation to a particular set of convictions about how the world is, and (c) an event is actually surprising only if it is noticed by the holder of that particular set of convictions.

Each location on the globe has its own set of climate-related hazards to cope with: floods, droughts, high winds, fires, tornadoes, hurricanes, tidal waves, ice storms, hail, frosts, and so on. While those who live in a particular location may be aware of their local hazards in a general way, they do not necessarily know the specifics about the risks associated with those hazards. Because they cannot reliably predict when the potential hazards will recur or how intense they might be, they are surprised to some extent when they do occur. While they may not have been directly affected by a particular hazard such as a tornado or a drought or a hail storm, they are not surprised when they read about them affecting other people – elsewhere. But when that same hazard directly affects them, they are truly surprised.

Thus there are knowable as well as unknowable surprises. The latter type (surprises that are beyond the realm of expectation given existing knowledge) is the traditional type that comes to mind to a person on the street. But a close look at what is surprising, and why, suggests that we might be able to "get ahead of the surprise curve". Norman Myers (1995) referred to knowable ("anticipatable") surprises, surprises that can be anticipated with some serious forethought. For example, people who live in an area that is susceptible to flooding would acknowledge that as a fact. No surprise here. What they do not know, however, is when that flooding will take place or how bad it will be. When it does happen, most people will be surprised. The same can be said of a range of hazards such as hurricanes or volcanic eruptions. They are knowable and are, to some degree, expectable. Yet, when they occur, people are surprised.

With respect to human responses to the same event, there are varying degrees of surprise. The following examples were taken from the scientific literature: people can be hardly surprised, mildly surprised, somewhat surprised, very surprised, extremely surprised, totally surprised. Myers (1995) introduced the strange notion of "semi-surprised". Thus surprise is best described in "fuzzy" terms, with the degree of surprise dependent on many intervening factors, such as personal experience, core beliefs, expectations, knowledge about a phenomenon or about the environmental setting in a region.

Change is a part of Nature and is to be expected. Yet, people tend to think in terms of persistence in the sense that today is like yesterday and tomorrow will be like today. People also tend to discount the past; that is, they give less importance to events that happened a few decades ago than to events of the recent past or to those taking place now. They discount the future in the same way. So, the probability of occurrence of a tornado or a famine or an anomalous natural hazard is set very low in one's mind because such events have not been experienced recently. In addition, this perception is reinforced by an optimistic belief that societies have learned from their past experiences or that new technologies can protect them from the impacts of potentially surprising climate-related events. Nevertheless, we continue to be surprised. The notion of climate-related surprise applies to El Niño events as well as other natural hazards.

## El Niño surprises

A discussion of El Niño surprises best begins with a focus on the early 1980s. Before then, there was little concerted effort to forecast El Niño or its worldwide impacts. So, let us start with the 1982–83 El Niño, which surprised the scientific community. As noted earlier, El Niño researchers, for the most part, were surprised by the timing of its onset, intensity, and the severity of its ecological and societal impacts. A story exists in the El Niño research community about a researcher who visited Peru in August 1982. At that time, he advised Peruvians that no El Niño would develop that year. His forecast was based on the fact that several precursors that he had associated with El Niño were not apparent. As his plane lifted off the tarmac at the airport in Lima, Peru, warm water was beginning to appear off the Peruvian coast. At a gathering of El Niño experts a few months later in October 1982 at Princeton University, the view of the participants based on existing "evidence" was also that an El Niño at that time was unlikely. We now know, with all the benefits of hindsight, that one of the strongest El Niño events of the century was in progress as they were meeting.

Now, at the start of the twenty-first century, the scientific community believes that it understands the physical aspects of El Niño well enough to forecast it reliably several months in advance of its onset. The record is quite clear that the science of El Niño has improved considerably since 1982. Some researchers forecasted an El Niño onset for 1990. However, Cane and Zebiak's model output did not support their forecast. The Cane and Zebiak model did, however, forecast an event to begin in 1991. Cane and Zebiak suggested that it would end in 1992. An El Niño began, as they and a few other research organizations had predicted, but the event did not end in November 1992, and in 1993 the surface waters in the eastern equatorial Pacific began to heat up again – unexpectedly. The 1993 event decayed, only to have yet another El Niño event re-emerge in 1994. This multi-year sequence of heating and cooling (but only around average) of sea surface temperatures across the equatorial Pacific surprised many researchers.

The unexpected behavior of the El Niño event(s) in the 1991–95 period surprised researchers as well as the public at different times and in different ways. For example, usually during an El Niño event the Peruvian anchoveta fishery off the coast of Peru is much less productive. However, when US forecasters claimed that a severe and lengthy El Niño was in progress in this period, it confused Peruvian fisheries managers whose fishing boats were catching record numbers of anchovy off the Peruvian coast.

## Examples of El Niño surprises

In mid-1997, I asked some researchers what might surprise them with regard to El Niño. I did this to gain a glimpse of "possible futures" that may be feasible from a geophysical standpoint (i.e., ways in which an El Niño event might develop). For example, a researcher suggested that the following occurrences would be surprising: an extended 1997–98 warm or cold event; a major warming or cooling that lasts only a very short time; a really huge event with sea surface temperatures (SSTs) in the Niño3 region going up to 30°C; an unprecedented El Niño-like episode in either the Indian or the Atlantic Ocean; and a major event occurring in the absence of a prediction of the event (as happened in 1982–83).

Another brainstorming session held in June 1997 with some ENSO researchers generated the following list of possible surprises. Although these are much more specific in detail, they are only meant to be suggestive.

- three years running of 1.5 degrees Celsius or more of an SST anomaly
- three years running of −1.5 degrees Celsius or more of an SST anomaly
- an SST increase of greater than 1 degree Celsius change in one monthly mean
- an SST increase of greater than 1 degree Celsius in one month
- the occurrence again of a perfectly classical (canonical) event
- an ENSO event that does not involve SST changes at the dateline or the equator
- a thermocline that is in a normal position and a warming that begins at the surface and grows downward (as in 1993)
- to have an ENSO event where the tropical Pacific SSTs and the Southern Oscillation Index go in opposite directions (winter of 1996–97)

- three ENSO warm events in a 10-year period
- a return to the 1950–70s ENSO regime (fewer El Niño than La Niña events)
- a big El Niño event in the central Pacific and no heavy rainfall in northern Peru (i.e., in Piura)
- if the 1997–98 El Niño event is equal to or stronger than the 1982–83 event (oops!)
- if, during a mature moderate ENSO event (by the end of the Northern Hemisphere winter: March), there was a sizeable area of below-normal SSTs in the northern Pacific around 140–170° W and 35–45° N
- to have a substantial El Niño in which, during the mature phase (January–February) there is no positive SST anomaly just south of Hudson's Bay and north of the Great Lakes
- and so on

It may be a useful exercise to continue to identify feasible, but not yet seen, aspects of El Niño-related behavior. In fact, some of the surprises suggested in June 1997 actually did occur during the 1997–98 El Niño and the transition to the 1998–2000 La Niña.

The statements that follow were taken directly from scientific papers. They, too, suggest varying degrees of surprise related to El Niño events. They make reference to the different sources of surprise, i.e., intensity, timing of onset, severity of impacts. Unlike the list cited above, these statements of El Niño surprise were based not on speculation but on scientific comments.

### *Some surprise related to El Niño's timing*

> Unlike its predecessors during the previous three decades, the 1982–83 warm episode was not preceded by a prolonged "buildup phase" with strong winds along the equator, and it did not exhibit what had come to be viewed as the typical "onset phase" around April.
>
> (Rasmusson and Wallace, 1983)

> The strongest El Niño episode in this century, in 1982–83, had a unique pattern in terms of the sequence of the warming and the time of the onset.
>
> (Kahya and Dracup, 1993, p. 2491)

> In direct contradiction to the canonical model, the timing of the SST anomalies was offset by a half-year off South America.
>
> (Enfield, 1989, p. 167)

> However, the perverse ENSO of 1982–83 developed in a quite different manner. It did not clearly show its hand until June 1982.
>
> (Rasmusson, 1984b, p. 10)

> Scientists were caught off guard by the unusual timing of the onset of the 1982–83 episode relative to the climatological mean cycle.
>
> (Rasmusson and Wallace, 1983, p. 1198)

### *Some surprise related to El Niño's magnitude*

> The 1982–83 warm episode is of particular interest because of its unusual intensity and evolution.
>
> (Gill and Rasmusson, 1983, p. 229)

> The 1972 event was remarkable in that maximum anomalies exceeded 4 deg.C.
>
> (Wooster and Guillen, 1974, p. 402)

The 1997–98 El Niño provides a yet-to-be-undertaken case study in surprise, because several aspects of El Niño-related changes in 1997–98 in the tropical Pacific were unexpected. For example, the SSTs rose at an unexpected rate in February 1997, and both the rate and timing were surprising; the intensity of the 1997–98 event confused scientists, as the intensity of this event helped to displace the 1982–83 El Niño as the "El Niño of the century"; although the end of the event was expected, the rapid drop in SSTs in May and June 1998 was not anticipated (about 1 degree Celsius per week). An intense La Niña event was forecast to develop immediately because of the unprecedented rapid decay of El Niño. However, it did not appear when expected in mid-1998, but emerged later in the year, was not an intense event, and lasted much longer than was originally forecast. In mid-1999, forecasters believed it would last into the spring of 2000; and so it did.

### *Other El Niño surprises*

Other surprising aspects of El Niño have been identified by scientists: "the unusual evolution of the 1982–83 event" (Gill and Rasmusson, 1983, p. 229); "the 1993 warm episode occurred against the backgound of a very anomalously shallow west Pacific thermocline" (Kessler and McPhaden, 1995); "the intensity of the wind collapse in the east was unprecedented" (Enfield, 1989, p. 167); "the pattern of oceanic anomalies and the climatic phenomena produced by the teleconnections turned out to be radically different from those observed on other occasions" (Enfield, 1989, p. 170); "the reverse order of the SST anomalies . . . which occurred first in the equatorial region and later at the coast" (Enfield, 1989, p. 173); "a precipitous fall in the Southern Oscillation pressure seesaw that was to last for 9 months" (Rasmusson, 1984b, p. 17); "rather abrupt onset" [of the 1982–83 event] (Harrison and Cane, 1984, p. 21); "weak occurrences of the phenomenon have remained unnoticed" (Wyrtki *et al.*, 1976, p. 346). The absence of a strong cold event (i.e., La Niña) from the mid 1970s until 1988 was also very surprising to El Niño researchers.

## Why are we still surprised by El Niño?

There are several reasons why researchers, policymakers, and the public continue to be surprised by almost every recent El Niño event. First, we have only recently begun to focus on El Niño as an important physical, biological, and social science research topic. Interest in the phenomenon evolved from concern at the end of the 1800s about how an El Niño event would affect Peru's guano birds and guano production (a fertilizer) through concern in the 1970s about international commodity trade (fishmeal and soybeans) and climate forecasting potential, to concern about its impacts on the frequency and intensity of weather extremes around the globe in the 1980s. With each successive event, researchers improved their understanding of the phenomenon, identifying new aspects as well as new "unknowns". However, scientists have experienced neither a full range of the ways that El Niño can develop and run its course nor all of its worldwide impacts. Even though the TOGA-TAO Array monitoring system was fully in place by 1995, well in time for the 1997–98 event, a forecast several months in advance of its onset or of its intensity was still not possible.

Recent press releases sported such headlines as "The Big Models Finally Got It Right" (Kerr, 1998) or "Climate and El Niño Easier to Predict Than Thought" (NSF, 1998). The gap between what we think we know about El Niño and what remains to be known is still quite large. Because of such news headlines and attempts by some forecasters to take credit for forecasting the onset of the 1997–98 El Niño (i.e., spin doctoring), we (the public, policymakers, researchers in other fields, and the media) tend to believe that scientists understand El Niño's behavior. Hence we have been unwittingly set up to be surprised by the next event, which is likely to differ in unexpected ways from its predecessors.

Usually, the public first hears about an El Niño after a long period of silence on the topic, when an event has been forecast. The media report on its stage of development at the time and on forecasts from research groups about its future evolution. Clearly, until now, the focus of research has been on forecasting its onset. As a result, a disproportionate level of attention of the public and policymakers (i.e., all of whom are potential users of El Niño information, including forecasts) has focused on developing a reliable forecast of El Niño's onset. However, not enough attention has been directed toward forecasting other phases of the ENSO warm event–cold event process. La Niña events (the cold side of the ENSO cycle) also represent a neglected area of research, as have the teleconnections related to average conditions in the central part of the tropical Pacific.

Under pressure from budgetary processes that support El Niño-related research, groups are often required (some might say "expected") to show their progress in their scientific projects to their research funding agencies from one year to the next or one event to the next. As a result, scientists are often put in the position of having to show progress on unrealistic time scales dictated by the annual national budget preparation cycles. Scientists are also under pressure, and perhaps in some cases even encouraged, to state their findings more quickly and in more positive terms than their research findings warrant. Each budget cycle pushes up the "ante", and agencies and scientists are under increasing pressure to provide more positive statements, if not concrete signs, to policymakers about progress in the societal value of their El Niño-related research. If one accepts this argument, then funding agencies bear some responsibility for future El Niño-related surprises.

In general, the public's contact with El Niño is via the media – television, newspapers, the Internet, and magazines. It is led to believe, like policymakers, that researchers have a better understanding and a better forecast capability than actually exists. By failing to acknowledge to the public that the El Niño mystery is not yet solved, the media are constantly setting the public up to be surprised when future events evolve differently from those of the past. The media compete over reporting on El Niño and its impacts and, as a result of such intense competition, perceptions of what El Niño is and what its impacts might be become greatly sensationalized and distorted.

Clearly, El Niño researchers have made considerable progress since the late 1960s when Jacob Bjerknes clearly described the linkage between equatorial oceanic processes and atmospheric circulation in the tropical Pacific. While researchers are learning more about the ENSO cycle with the passing of each warm or cold event, it seems that the problem of ENSO grows bigger than was originally believed. With each event, researchers identify new areas of El Niño/La Niña research to pursue.

## Concluding comments

One thing the research community can do to reduce the likelihood of being surprised by each future El Niño or La Niña event is to implement a suggestion made by Thompson and his colleagues about surprises in general. They suggested that we "collect our surprises (as if they were botanical specimens) and scrutinize them for their similarities and differences" (Thompson *et al.*, 1990, p. 52). This unique suggestion is not so bizarre. We could, for example, go back to the major El Niño of 1891 to identify the state of knowledge at that time and then attempt to identify the surprising aspects of that particular event. We can do this for each successive major event, noting new surprises and how and why they arose. We can also identify why some of the earlier surprising aspects of El Niño are no longer considered to be surprising. This retrospective approach could help to identify "expectable" surprises, enabling societies to prepare better for them. It would reduce some of the potential surprises generated by future El Niño and La Niña events or by their environmental and societal teleconnections.

Bumper stickers adorn cars in America, carrying slogans, advertisements and, sometimes, thought-provoking ideas. One that recently captured my attention and made me think more seriously about El Niño-related research was the following: "Don't believe everything you think."

# 15 What people need to know about El Niño

## Introduction

At the end of the twentieth century, El Niño was in the news . . . big time. Although there had been a major El Niño in 1982–83, it seems that people on the street as well as policymakers had forgotten about it. There had also been an El Niño episode in 1986–87 and, again, a set of them in the first half of the 1990s, but from a global as opposed to a national perspective they were viewed as having been rather moderate in their impacts on societies. The US media, however, took the forecasts of the 1997–98 El Niño much more seriously than they had for any previous El Niño. Apparently, the media worldwide paid attention to El Niño once scientists in Australia, the USA, Peru, and Ecuador talked about it and its potential impacts in distant places such as Australia and South Africa.

The media coverage from late summer 1997 to mid-1998 of the 1997–98 event has proven to be a mixture of fact and speculation about the causes and consequences of El Niño. While the media searched for good stories to tell the public, scientists were not reluctant to share their views, however speculative, with the media. Stories of El Niño appeared daily in the print and electronic media. In addition, just about every strange weather-related event was blamed on El Niño. On the Internet, El Niño chat groups were established. One interesting chat group comment was the following: "El Niño is a weatherman's holiday. If it rains, it's El Niño. If it's a drought, it's El Niño. The weatherman doesn't have to analyze anything. He just has to blame El Niño!".

In response to the way that El Niño facts were being mixed together with speculation about El Niño and its impacts and the way they were presented to the public (sparking the allegation that the media were "hyping" El Niño), it became obvious to me that there were at least seven basic points about El Niño that people ought to know in their attempts to separate fact from speculation. In addition, there are nine El Niño-related traps of which the public should be aware. I hope basic statements such as these will enable the public to better calibrate El Niño- and La Niña-related information that they hear in media stories and from scientists.

## Seven things that people ought to know about El Niño

### *1. El Niño does not represent unusual behavior of the global climate*

El Niño is usually described as a climate anomaly, or as an unusual or abnormal interaction between the air and the sea in the central and eastern equatorial Pacific Ocean. As such, it has been viewed as not being a part of the normal climate system. In fact, El Niño *is* a normal part of the climate system and not separate from it. While we can talk about how the sea surface temperatures in the eastern and central equatorial Pacific Ocean may depart from some mathematically defined average condition, we must not view that departure as abnormal. El Niño (a warm event), like its counterpart La Niña (a cold event), is an integral part of the global climate system. Making this distinction more obvious and explicit can help people to realize that El Niño events have occurred for thousands of years and that they are to be expected and, hence, prepared for. Indeed, to go through a decade or two without an El Niño would be truly unusual. However, El Niño events can be extraordinary in their extreme level of intensity and that would represent unusual behavior of the phenomenon when compared to its historical record.

### *2. El Niño is part of a cycle*

El Niño gets all of the attention, not only from the media but from physical and social scientific researchers as well. But, it is important to remember that El Niño is the warm phase of a cycle that also includes a cold phase, referred to as La Niña. There has been less interest in La Niña over the past two decades, because there have been far fewer cold events than warm ones in that period. Nevertheless, extreme weather events around the globe have been associated with La Niña. Scientists say that La Niña-related extreme events are the opposite of those caused by or related to El Niño; for example, drought usually accompanies El Niño in Southern Africa, while La Niña is associated with flooding in the region. Researchers are now focusing some of their attention on the cold part of the cycle. Statements about symmetry between El Niño and La Niña await validation by the research community.

Policymakers, government agencies, physical scientists, social scientists, and the public are increasingly focusing on El Niño as one of the few bright spots in attempts to forecast future states of the atmosphere and their impacts on societal activities. There will still be some failures (i.e., misses) in the forecasting of future El Niño events, but scientists are increasingly developing a more complete understanding of this important natural phenomenon. Combining this increased knowledge with an improved understanding of how La Niña can affect weather around the globe will in time enable governments and people to cope with the adverse weather anomalies that they spawn.

### *3. Every weather anomaly throughout the world that occurs during an El Niño year is not caused by that El Niño*

We must be careful which adverse impacts on societies and on ecosystems we blame on an El Niño event. There is a tendency to blame just about everything that happens during an El Niño episode (which can last 12–18 months) on that particular El Niño. However, only some parts of the globe are directly influenced by El Niño-spawned regional climate anomalies, and even those areas are not necessarily influenced in the same way by different El Niño events. Every year, even in non-El Niño years, extreme record-setting weather events are occurring at various locations around the globe. The linkages between El Niño and regional climate anomalies have been identified through: (a) observations of direct linkages between warm surface water in the equatorial Pacific and distant regional anomalies (such as drought in Papua New Guinea or Australia); (b) statistical measures identifying probable linkage (such as droughts in Mozambique or Ethiopia); and (c) wishful thinking, whereby people believe that a particularly disruptive weather impact was due to El Niño, even in the absence of evidence to support their belief.

### *4. El Niño has a positive side as well*

Most studies do not focus on any of the positive aspects of El Niño. Yet, there are situations where El Niño benefits local ecosystems and societies. For example, during an El Niño the number of hurricanes along the Atlantic and Gulf coasts of the USA are greatly reduced in number (Pielke and Landsea, 1999). During the 1997 El Niño year we did not have a devastating blockbuster hurricane or any major damage from tropical storms. In fact, it was an unusually quiet hurricane season. As another example, during an El Niño period there is a sharp increase in the amount of wild shrimp larvae off the coast of Ecuador. This is good for that country's shrimp industry, although it is not so good for the privately owned shrimp larvae hatcheries. Very little effort has been focused so far on compiling the instances where societies have benefited from El Niño's occurrence. In fact there has been virtually no sustained research on what it might mean for a society or an economic sector to benefit from an El Niño event.

### *5. There will continue to be surprises associated with future El Niño events*

Scientists have really only concentrated their research efforts on El Niño as a Pacific basin-wide phenomenon since the mid to late 1970s. They have not yet witnessed all of the ways they can form; nor have they witnessed all of the combinations of ways that they can simultaneously affect societies and ecosystems worldwide. Thus each succeeding El Niño will probably surprise scientists as well as the public in its timing of onset, frequency or in the magnitude of its impacts (e.g., level of destruction or misery).

### *6. The impact of global warming on El Niño is not as yet known, speculation notwithstanding*

Despite the increasing speculation about the possible ways that global warming of the atmosphere could affect El Niño events (timing, frequency, magnitude, duration), the scientific community is unable at this time to say with any degree of confidence what the impacts of a global warming will be on El Niño, media speculation and scientific "guess-timates" notwithstanding.

### *7. Forecasting El Niño is different from forecasting the impacts of El Niño*

Scientists are trying to forecast El Niño by focusing their research efforts on identifying those characteristics of El Niño that appear early in its onset and development phases. However, the success (or failure) to forecast an El Niño several months in advance of its onset is different from forecasting the impacts of that particular El Niño that occur months later. Forecasting impacts on societies around the globe requires different research methods. Each El Niño seems to cause a different set of impacts (combination of droughts, floods, fires). However, some impacts in some specific locations tend to recur during most El Niño events (e.g., the anchoveta population off the coast of Peru declines, drought occurs in northeastern Australia and in Indonesia, the number of Atlantic hurricanes declines, etc.). In 1997 scientists were unable to forecast El Niño's onset in

February 1997 but were able to produce successful forecasts of El Niño's impacts months later in, for example, northern Peru, southern Ecuador, North America, Indonesia and the Philippines. Problems (and success rates) in forecasting El Niño (the event), therefore, are different from those related to forecasting El Niño's impacts.

## Nine El Niño "traps" that people ought to know about

### *1. Scientists do not agree on the list of El Niño years*

There is no single list of El Niño years that has been universally accepted by researchers. As a result, various researchers include different years in their own lists of El Niño, La Niña, or normal years. For example, consider the 1972–73 El Niño event. Some researchers refer to it as the 1972–73 event. Others suggest that 1972 was an El Niño year, and that 1973 was a cold event (La Niña) year. Which one is correct? Look also at the 1982–83 El Niño. Again, some researchers consider both years as El Niño years, while others have considered 1982 normal, and 1983 as El Niño. Most recently, we have referred to 1997–98 as El Niño years, with 1997 as El Niño year (0) and 1998 as El Niño year ( + 1). Yet, 1998 was also La Niña year (0). Thus, there are various combinations of years that could be used in an assessment to get just about any correlation one might desire. This is troublesome for those interested in objectively determining statistical correlations (relationships), or absence thereof, between El Niño events and crop production, streamflow, disease and pest outbreaks, extreme weather events in distant locations (called teleconnections), and so on. It is important for the users of El Niño information, including teleconnection correlations, to identify whose list of years they rely on to determine a specific correlation. How reliable, then, are the various correlations we hear about with regard to El Niño or to La Niña? It is imperative that the El Niño research community develop a commonly accepted list of El Niño and La Niña events.

### *2. Forecasting El Niño's onset does not tell us much about its other characteristics (e.g., intensity, frequency, duration)*

The lion's share of research funding and effort has been focused on forecasting the onset of El Niño. There has been relatively less interest in research on other El Niño characteristics such as its intensity, frequency, and duration, as well as the location of its impacts on societies and on managed and unmanaged ecosystems. However, the phenomenon goes through various phase changes, e.g., the onset, growth, mature, and decay phases. Unfortunately, forecasting its onset several months in advance does not tell us much about the characteristics of El Niño's other phases. Much more research is therefore needed on aspects of El Niño other than just its onset.

### *3. Monitoring El Niño is different from forecasting it*

Because El Niño is better monitored than ever before, this does not mean that it is being forecast better than ever. A systematic monitoring capability was set up in 1985, following the establishment of the decade-long TOGA program from 1985 to 1994. Researchers are still working hard to develop *the* best model of air–sea interaction in the equatorial Pacific Ocean. Until a few years ago, the best model was the one developed in the mid-1980s by Mark Cane and Stephen Zebiak of the Lamont–Doherty Earth Observatory (Columbia University). As late as the mid-1990s, their model was considered successful and a flagship model for forecasting El Niño, having been used to forecast warm and cold events correctly between 1986 and 1992. Since 1993 the model has apparently not fared well. For example, most recently the model produced a forecast of a cold event for 1997–98. As we now know, the 1997–98 event turned out to be a prime contender (with the 1982–83 event) for the label of the biggest El Niño of the twentieth century. Only time and a run of successful forecasts by a model (or models) will determine whether a new flagship model for producing reliable El Niño forecasts has been developed.

### *4. When viewed as an* event, *El Niño evokes different concerns than when it is viewed as part of a* process

Many people have come to view El Niño as a discrete event. It starts at a certain point in time and ends several months later. Once an El Niño is said to be ending, interest in it and its impacts wanes drastically. The general public and policymakers alike relax, believing that they have nothing to fear (and in many cases, nothing to do) until the next El Niño event is forecast. The view that El Niño is a discrete event has been reinforced by the way it is covered by the media. With the decay of an El Niño, the media lose interest in it as "story material".

However, a warm event can last from 12 to 18 to 24 months and is really part of a longer cycle and larger process, the ENSO cycle and interannual climate variability, respectively. The ENSO cycle includes the cold phase and the normal phase, along with the warm phase. Viewing El Niño as a process should help to maintain a high level of interest in the phenomenon between the peaks of El Niño events and not just during the events. In this regard a scientific and a media focus on cold events (La Niña) would help to shift perceptions of El Niño from event to process. These must be considered as two different aspects of El Niño.

*5. We have not examined enough El Niño events to know all the ways they can develop and play out. The same applies to El Niño's impacts on societies and on ecosystems*

The truth of the matter is that scientists "discovered" El Niño as we now know it (a Pacific basin- wide phenomenon) only in the 1960s, when Professor Jacob Bjerknes described the linkage between oceanic processes in the eastern equatorial Pacific (sea surface temperature changes) and sea level pressure changes across the Pacific basin (the Southern Oscillation). Researchers began to focus on it more intensively after the 1972–73 event, and once again after the 1982–83 event sparked worldwide interest in the phenomenon. Monitoring El Niño began systematically in 1985 with TOGA. The point is that we have not examined, as yet, the various ways that an El Niño might develop. For example, the sea surface temperatures warmed first along the Peruvian coast in the 1972–73 event, in the central Pacific during the 1982–83, and in the eastern and western part of the equatorial Pacific during the 1997–98 event. In addition, we have not yet realized all of the various combinations of societal impacts that an El Niño can spawn. Do not be surprised when an El Niño does not "perform" according to our expectations.

*6. A pretty website does not El Niño expertise make*

Given the high level of media attention focused on the 1997–98 event, there has been an explosion of websites on the Internet, as well as of experts on El Niño. The Internet has become a major source of information on the phenomenon, and that information is free for the taking. Anyone who can access the World Wide Web can find information on its various aspects: news, graphics, movies, cartoons, "hype-watch," chat groups, and so forth. The Internet is an information highway, but how is one expected to sort out reliable information from the rest? In the absence of any way to screen the pieces of El Niño information one encounters on the Internet, the pressure is on the users of such information to develop ways in which to calibrate the information on which they choose to rely. The use of El Niño information taken from a website requires a "buyer beware" attitude if not a label. In other words, know *your* El Niño expert's strengths and weaknesses.

*7. The media do not have a neutral interest in reporting El Niño*

The media are not neutral when it comes to reporting on El Niño. They search for headlines that can grab the public's attention. They are driven by a daily search for a bigger share of the readership market. The waves of media interest in El Niño (based on my personal encounters from July 1997 to spring 1998) were as follows.

1. *What is El Niño?* This first wave comprises reporters and editors interested in the fact that the scientific community is calling for an El Niño event to take place.
2. *Again, What is El Niño?* The second wave is made up of reporters and editors who missed being in the first wave of interest in El Niño stories. They said that they were asked (if not ordered) by their editors to do a story on El Niño, like the ones that appeared in the first wave of media interest.
3. *What's wrong with the science?* In the third wave of media attention, reporters asked this question, based on the feeling that some of the forecasts of impacts were wrong, or that the El Niño was developing in unexpected ways. They wanted to seek out negative comments on the state of the science of El Niño forecasting.
4. *What is not being said about El Niño that we can say?* The media then sought to find aspects of El Niño that the others reporters and editors had missed.
5. *El Niño hype?* The final wave of queries from the media was focused on "hype". Too much media coverage was just media hype. Interestingly, many of the reporters and editors who were in the first wave of interest in writing about El Niño were the ones who wrote stories about the eventual overexploitation of El Niño.

### *8 Beware of the use of El Niño analogies (e.g., this event is like the 1972–73 or 1982–83 event)*

At the outset of the 1997–98 event, there was talk by some researchers that it was like either the 1972–73 or the 1982–83 event. Once the media picked up on 1982–83, they ran with it and scientists stopped referring to the similarities with the 1972–73 event. The media had film footage on the event of the early 1980s, but most likely did not have such footage for the event of the early 1970s. Once the specter of a return of an El Niño the size of that in the early 1980s, the notion of the "El Niño of the century" was invoked, people began their search for possible impacts by focusing on the same locations that were adversely affected by the 1982–83 El Niño.

There have been other warm events in the twentieth century, but we do not have reliable information on events other than the most recent ones. Even picking the wrong analog year, when an El Niño has been forecast, can have major implications for selecting the appropriate prevention, mitigation, and adaptation strategies for affected regions around the globe.

### *9. It is very tricky, risky, and potentially misleading to blame any specific weather event on El Niño*

During the usual duration of an El Niño (12–18 months), many anomalous weather- and climate-related events take place. A large number of these are likely to be blamed on El Niño. In some cases, that may be valid; for example, the storm fronts that hit the coast of California with great regularity during the 1997–98 event. Other extreme weather events that occurred during that El Niño, such as the killer tornadoes in Florida in February 1998 or the ice storms in Quebec in January 1998, do not have clear and definitive causal links to El Niño. Although such occurrences may be plausibly linked to El Niño or El Niño may have played a contributory role, it would be inappropriate to blame them solely on El Niño. Research on linkages of specific weather episodes to El Niño or La Niña is under way (Barsugli *et al.*, 1999).

In a general sense, one could argue that El Niño *can* take the blame. If, for example, any storm that occurs during winter can be considered a winter storm, then any storm that occurs during a winter in which an El Niño is taking place can be considered an El Niño-related storm. But this is not necessarily a useful attribution when it comes to the forecasting of, or societal preparations for, El Niño's impacts. El Niño affects seasonal characteristics and therefore should not be blamed as the main cause of specific weather events and anomalies within that season.

# 16 Usable science

El Niño events affect the lives of hundreds of millions to billions of people, either positively or negatively, either directly or indirectly, and most directly in the tropical countries that girdle the globe. Any early warning, "heads-up" information about the impending or the actual beginning of an event can be used by those in power to mitigate its possible adverse impacts and, where possible, take advantage of its positive effects.

As far as anyone can tell, El Niño events are here to stay. Societies must learn more about them and learn how to use El Niño-related information to their advantage. The same advice applies to cold events. A society that is forewarned about El Niño or La Niña will be better prepared to cope with its effects.

The research community dealing with El Niño has for the most part focused on forecasting the onset of the event. While this is very important and useful to societies potentially affected by it, it is only part of the picture. If all El Niño-related research were to be halted today, there would still be considerable value in using existing information about El Niño – it would be useful for decisionmakers in many countries. National examples of the benefits of improved *awareness* of El Niño are increasing, especially as a result of the 1997–98 event, in Ethiopia, Southern Africa, Kenya, Canada, USA, Australia, Vietnam, Fiji and other Pacific island nations, Ecuador, Peru, Brazil, Chile, Costa Rica, and Cuba.

Figuring out the science of El Niño is not unlike trying to piece together a jigsaw puzzle. Some pieces are easy to identify – those with the flat edges that form the outside border of the puzzle. Others are identified by the pictures on the puzzle, and they are pieced together forming isolated clusters that by themselves provide a set of autonomous, unconnected pictures. The clusters are not yet attachable to each other or to the outside frame.

With regard to the jigsaw puzzle analogy, there are many pieces whose shapes and colors are not readily identifiable and, therefore, become subjected to trial and error. This is where we are at present with El Niño research. To date, El Niño researchers have identified the framework and several clusters of knowledge centered on characteristics of the El Niño phenomenon, such as Kelvin waves, teleconnections, the winds, the warm pool, the Southern Oscillation. The hard part lies ahead. However, there is much to be learned before we can say that we have reduced scientific uncertainty enough to make forecasts of El Niño's onset consistently reliable. The recent El Niño event(s) in the 1991–95 period and the unexpectedly rapid development and decay of the 1997–98 event suggest that we do not have all of the key El Niño puzzle pieces on the table in front of us.

The physical and biological research communities are poised to continue their efforts to unlock the remaining mysteries of air–sea interaction in the Pacific Ocean and their teleconnections to weather and climate anomalies. The social sciences and public health specialists are involved. At the beginning of the twenty-first century, the El Niño research community includes several social scientists who seek to identify ways to apply research findings to the needs of decisionmakers in various sectors of society. Clearly, El Niño research is one of the bright spots on the scientific horizon of the new century. This relatively neglected natural environmental change with implications for the globe has clearly begun to get the respect it deserves.

We do not know for sure how the global climate will change significantly in future decades. Clearly, we are only beginning to gather clues about how global warming might affect El Niño's characteristics. If, however, governments want to gain an idea of how well prepared societies might be with regard to coping with the consequences of climate change, they can support assessments of how well societies cope with climate variability and extreme events at present. Numerous studies now exist about how different groups and nations have coped with floods, droughts, hurricanes, frosts, fires etc. These are often treated as random events. However, El Niño is a recurring event, a stimulus–response experiment of sorts in the Pacific, which can provide researchers every few years with information about how well (or how badly) societies cope.

Realizing the importance of knowing more about El Niño (rather late in the game, I might add), the UN General Assembly passed a resolution in November 1997 calling on its agencies to review the scientific and technical state of knowledge about the ENSO cycle (the WMO became the lead UN agency for this review) and about the disaster aspects of the 1997–98 El Niño (the International Decade for Natural Disaster Reduction (IDNDR) became the lead agency for this part of the review). Until this time, however, the IDNDR had not been concerned with El Niño or its societal aspects. The government of Ecuador invited the WMO and the IDNDR to convene a conference to share their findings. This conference was held in Guayaquil, Ecuador, in November 1998. Final reports were issued by the agencies (WMO, 1999; IDNDR, 2000).

In mid-1999, the United Nations Environment Programme (UNEP) and the US National Center for Atmospheric Research (NCAR) received support to study the 1997–98 El Niño from the UN Foundation for International Partnerships (UNFIP), a new international non-governmental organization created by a gift of $1 billion from US media mogul Ted Turner to the UN. People no longer see El Niño as something that happens every once in a while off the western coast of South America. Many policymakers and much of the public have moved beyond that image of El Niño.

Any improvement in our understanding of the event, its teleconnections and its societal and environmental impacts can be of value. Society has nothing at all to lose with regard to continued, if not increased, support for El Niño-related physical, biological, and social science research. It is important for the public – not only policymakers – to know about El Niño. In fact, El Niño research presents society with a "win–win" situation.

Society has not paid as much attention to, or put resources toward, the use of knowledge as it has put toward the generation of that knowledge – particularly during the Cold War decades, when the production of scientific knowledge flourished. It is time for society to improve its efforts to better connect the production of scientific information to its use by society. It has become more important than ever to bridge the gap between scientific output and societal needs. We, as a society (individuals as well as government funding agencies), must put more effort and resources into mining the scientific information that is being produced in ever-increasing quantities for use in addressing societal needs and wants. El Niño research output provides a good contemporary example of usable scientific information.

El Niño is now acknowledged to be a naturally recurrent phenomenon that disrupts human activities, for good and ill, in many parts of the globe. Social scientists now show considerable interest in El Niño and its impacts on ecosystems and on economies as can be seen from an increasing number of social assessments (e.g., Fagan, 1999; Orlove and Tosteson, 1999; Zapata-Velasco and Sueiro, 1999). For example, in a letter to the British medical journal the *Lancet*, Dutch representatives of Medécins Sans Frontières, Bouma and colleagues, suggested that El Niño has been the driving force behind periodic malaria epidemics. They went on to suggest that

> the advances made in the past decade in meteorological forecasting of the phases of the Southern Oscillation (warm and cold events) may help to predict areas at risk of malaria epidemics . . . This offers possibilities for developing early warning systems that can facilitate epidemic preparedness.
>
> (Bouma *et al.*, 1994, p. 1140)

El Niño is no longer just a physical scientific curiosity.

## By when should scientific research be applied to be considered usable?

Some people recognize the need to support basic research that may have no immediate, readily identifiable societal benefits. Others put a higher priority on instant gratification and support only what they consider to be applied science. Still others focus on what has been referred to as "curiosity-driven" research. El Niño research efforts must include each of these elements.

The potential socioeconomic payoff of improved knowledge of El Niño and La Niña is great for developing and industrialized countries alike (but for different reasons). Thus El Niño research efforts demand continued moral and financial support from governments worldwide. Those payoffs are likely to be incremental. Policymakers should recognize this incrementalism as a highly probable scenario. For its part, the scientific community must avoid generating misperceptions through "spin doctoring" about what it can offer in the near term in the way of usable El Niño and El Niño-related information.

The following few pages were prepared in late 1999 by Peru's president, Alberto Fujimori, who had used the information in the 1997–98 forecast of El Niño to take societal precautions. He provides a personal account of the event and his government's response to the forecast and to El Niño's potential and actual impacts. His account provides a perspective seldom seen in print: that of a government leader holding the highest political office whose country's citizens and economy are plagued by a recurrent hazard-spawning natural phenomenon.

# A president's perspective on El Niño

ALBERTO FUJIMORI *of the Republic of Peru*

By the end of July 1997 we, in Peru, were aware that a powerful El Niño would very probably occur toward the end of that year or the beginning of 1998. Unlike on previous occasions, we were able, for the first time, to put in place a successful anti-El Niño strategy. Once scientists in Peru and abroad had predicted that there would be an El Niño of biblical proportions, we were able to implement, quickly and efficiently, a plan to prevent as much disaster as possible.

Normally El Niño manifests itself as the warming of sea surface waters in the tropical Pacific and the weakening, even the reversal, of the trade winds over these waters. The arrival of the warm waters at the coast of Peru is one of the most spectacular aspects of global climate variability on an interannual time scale.

No other port in the world exhibits variations in sea surface temperatures and subsurface temperatures as extreme as those of our ports of Paita, San Jose and Chicama: as much as 10 degrees Celsius, measured at the same hour and day, from one year to the next. These variations have profound impacts on all aspects of coastal life in Peru each time they occur. From the Pre-Columbian epoch until today, every time the warm waters bathe the coast of Peru the activities of the Peruvian people are acutely affected. Our archaeologists have determined that the vanishing of entire cultures of ancient Peru can be linked to the impacts of El Niño on their food production.

In 1972 a large-scale warm event and our limited understanding of it devastated our fishing industry, the largest in the world at the time. It took many years for it to recover. The impact of the 1972 event served to awaken the interest of scientists such as Dr Michael H. Glantz – an old friend of Peruvian scientific institutions – in this particular climatic and oceanographic phenomenon.

The next extraordinary El Niño appeared in 1982–83 and had an enormous effect on our country, resulting in the loss of life and property. It took years and thousands of millions of dollars to rebuild the infrastructure and the economy of the affected regions. Even a 10% cut in salaries had to be imposed in order to contribute to the reconstruction efforts.

In 1997–98, for the first time in 175 years of our republic's history, the effects of El Niño, despite their magnitude, were not catastrophic. Six months before the onset of the event we took preventive measures. The government of Peru, as it has done when confronted with problems of many kinds, immediately laid out a strategy. The difference this time was that the strategy had to confront a natural event, something not under human control. Moreover, the forecasts of national and international experts called for an El Niño of unprecedented magnitude.

Our preventive strategy was not generally understood and to pursue it was politically very risky, because during the six months preceding this El Niño event it was necessary for the President of the Republic to devote a significant amount of his time to the management of the El Niño problem. Quick decisions had to be taken at the highest level in response to the national emergency. In a very short time, resources to sustain the infrastructure, machinery, medical, and sanitation needs of the vulnerable regions were put in place to enable them to confront an El Niño that was expected to be at least as intense as the 1982–83 event. This was done despite the reluctance of some forecasters to predict such an intense event, which by the end of 1997 was expected to begin in early 1998. Sometimes we worked 24 hours a day on the tasks of prevention and emergency response, including Christmas and New Year's Day.

There was a great risk that all the preventive measures would prove to be insufficient to mitigate the effects of El Niño, and consequently that the El Niño response strategy would fail. However, it was very clear to me that the strategic plan had to be carried out as decisively and rationally as possible to avoid huge floods in the northern cities and large-scale damage to the fields, i.e., the areas most susceptible to El Niño's impacts. Not to do so would have left the country at the mercy of Nature. The magnitude of potential impacts not only justified our measures but made it imperative to act not through "central planning" but for "planned centralization". This also created political difficulties, because local authorities wanted to confront El Niño's impacts by separately administering the resources at their disposal. That would have created chaos.

There were some instances of an internal lack of understanding, but there were other instances of gratifying comprehension. Mr Burkis, the World Bank's Regional Director for Latin America, accompanied me on an inspection trip and declared that Peru was the first country in the world to engage in this type of El Niño-related preventive work. After evaluating our efforts, the *Washington Post* referred to me as the Peruvian General Patton – a stimulating exaggeration.

The management of the ensuing crisis, which a CNN reporter called a catastrophe in slow motion, had put in place an effective plan that, for the first time in the history of Peru, prevented a drop in the Gross National Product during a year of severe El Niño. And this was also a period when all nations in the region had registered losses as part of a global recession.

The pro-active management of the expected crisis required budgetary support that had not been planned for yet was appropriately financed. We were able to acquire machinery and earmark resources not originally planned for by the budget office. The aims of the strategy that we pursued were as follows:

1. To guarantee sufficient stocks of food, clothing, and medicine to the isolated populations.
2. To use the excess water from the torrential rains for emergency afforestation projects. This was accomplished very successfully on the northern coast of Peru.

The 1997–98 El Niño was of such magnitude that the rivers of northern Peru reached "Amazonic" volumes. The containment of the volumes of water in the rivers that flow through cities such as Tumbes, Chimbote, and particularly Piura, is the best testimony of the success of our preventive strategies. Unlike during the 1982–83 El Niño, none of these cities was flooded, despite the river waters registering volumes that were 50% to 100% larger than the previous record maxima during the 1982–83 El Niño.

The result of the preventive measures was that flooding was avoided in towns and cities that had been demolished in 1983 by volumes of water that were only half those experienced in the 1997–98 El Niño. Of course, some homes were damaged, but there were neither casualties nor flooding of rivers such as La Leche in the Lambayeque district on the northern coast of Peru, despite the fact that La Leche flow had reached 1000 cubic meters per second. The bridges were enlarged and a 25-kilometer long canal was constructed to divert the excess water toward the desert. Without this canal to divert the water, many population centers would have been inundated and there would have been many fatalities. So much water was diverted to the desert that a lagoon was created and named Lake La Niña. It became the largest body of fresh water in Peru, after Lake Titicaca. Indeed, Lake La Niña became quite a tourist attraction.

In the city of Ica, 300 kilometers to the south of Lima, an extraordinary avalanche of rocks and mud disrupted basic services and buried entire neighborhoods. The number of those in need increased by 100 000 within 2 hours. To manage the emergency, a network of *tambos* (food piles like those used by the Incas but this time made using modern technology) was strategically located so as to be available to the affected populations, and a rescue system was put in place, including helicopters and launches.

Foreseeing the availability of rainwater in the desert, an emergency afforestation program was set in motion. Ten thousand kilograms of seeds of native forest species were collected: carob, sapota, faique, huaycan, hualtaco, among others. The carob is excellent as livestock feed and, as with the other tree species, is also used as timber. The faique, huaycan, and hualtaco are timber species that are almost extinct but are much sought after for parquet flooring. Sapota is used by artisans. Initially, the experts deemed it improbable that the goal of collecting a metric tonne of seeds could be reached. However, the enthusiasm of the local population surpassed all expectations and 130 000 kilograms were collected. These seeds were sown every 10 meters (with 10 seeds per hole) over 400 000 hectares. Today, more than 250 000 hectares that were formerly desert have been converted to forest.

There are several lessons to be learned from this El Niño:

1. With a 6-month lead time, strategies can be completed prior to the onset of El Niño's impacts.
2. It is possible to design and construct sufficiently strong, well-built bridges.
3. The areas and roads that are most vulnerable to floods can be identified and bridges and overpasses constructed.
4. It is possible to protect the cultural patrimonies that might be affected.
5. It is possible to cultivate the temporarily wet marginal areas in order to produce food crops that require a short growing season.
6. It is possible to continue to convert deserts into dry forests, using native species and with the participation of local populations.
7. It is possible to respond immediately to the occurrence of natural disasters and to try to maintain the normal lifestyle of the population.

In summary, with the possibility of further extraordinarily large El Niño events occurring in the future, the main infrastructure components, such as bridges, roads, and canals, all of which require large financial investments, should be designed taking into account the largest El Niño events that can be envisioned. Then, knowing a few months in advance the time of onset of an El Niño event or of its potential impacts, it is possible to pursue preventive measures, alert the population, protect the national patrimony, and take advantage of the availability of larger quantities of water in those regions where water is normally the limiting factor for agriculture, livestock rearing, and forestry.

This, then, is our experience of using the information provided by scientists such as Michael Glantz who have studied the impacts of climatic variability on various populations. My administration shares this effort to help future leaders to confront the biggest of all challenges: the challenge that Nature imposes on us. My words are thus a modest contribution to the scientific investigative work that deserves gratitude and respect.

# Appendix: Chronology of interest in El Niño

MICHAEL GLANTZ, *NCAR, and*
NEVILLE NICHOLLS, *Australia Bureau of Meteorology Research Centre*

**1876–77**: According to scientists today, an El Niño in this period was considered to be an intense one that was associated with famine in India. In retrospect, El Niño is now viewed as a contributor to Indian famines. This particular Indian famine sparked the scientific search for ways to forecast climate variability on the Indian subcontinent. Henry Blanford (the first Imperial Meteorological Reporter to the Government of India) noted high atmospheric pressure over the Indian region during the monsoon failure of 1877 and contacted other meteorological observatories for information about atmospheric pressure and rainfall.

**1885**: On the basis of Blanford's work, the India Meteorological Department began to issue forecasts of Indian summer monsoon rainfall.

**1888**: Australian Charles Todd, prompted by Blanford's earlier requests for data, noted a relationship between Indian and Australian droughts. This was the first evidence of teleconnected climate anomalies.

**1887–1903**: John Eliot, Blanford's successor, widened the scope of concern of his precursors by using, *inter alia*, Nile floods and southern Australian and South African data to forecast the Indian monsoon.

**1891**: A major El Niño negatively affected Peru, bringing heavy rains and flooding to northern Peru in April.

**1892**: Research papers referring to El Niño were presented at a conference of the Geographical Society of Lima (Peru).

**1894**: Peruvian geographer Victor Eguiguren published an article on the heavy rains in Peru (Eguiguren, 1894). (NB: "Rains" in this context are a surrogate for El Niño.)

**1895**: American geologist, Alfred F. Sears, wrote of the folklore in Peru about a climate phenomenon referred to as the "septennial rains". Such quasiperiodic rains are apparently those heavy rains associated with El Niño (Sears, 1895).

**1897**: Hugo Hildebrandsson reported on the out-of-phase relationship between sea level pressure at Sydney (Australia) and sea level pressure at Buenos Aires (Argentina) (Hildebrandsson 1897).

**1899**: The year of a major famine in India.

**1900 (June)**: Douglas Archibald reported on Indian monsoon forecasting in a US scientific journal (*Monthly Weather Review*). He noted that the Indian Meteorological Service forecasted a weak monsoon, if the deviation from normal of Indian region sea level atmospheric pressure was increasing prior to the monsoon onset, and a good monsoon if it was falling. This is a "symptom" of the yet-to-be-identified Southern Oscillation.

**1909**: Gilbert Walker (John Eliot's successor) issued the first forecast of the Indian monsoon using South American sea level pressure and other variables (similar to what we now call the Southern Oscillation).

**1910**: Walker and, independently, E. T. Quayle (in Australia) published statistical methods to forecast Australian rainfall using what we now call the Southern Oscillation.

**1910–37**: Walker documented the Southern Oscillation, using the term for the first time in a paper published in 1924.

**1920**: Coker's article about guano birds appeared in *National Geographic*, "Peru's wealth-producing birds". These birds eat fish (mainly anchoveta) and produce guano on the rocky islands along Peru's coast. Guano is an excellent fertilizer and any climate event that adversely affected its production (such as El Niño) was duly noted (Coker, 1920).

**1923**: R. C. Murphy wrote an article about El Niño and guano birds' mortality in the *Geographical Review* (Murphy, 1923).

**1925**: A strong El Niño event occurred, with major damages reported in Peru. Peruvians consider this to be one of the most notable of the twentieth century.

**1926 (January)**: Murphy wrote about the impacts of the 1925 El Niño in Peru (*Geographical Review*). Murphy's writings generated interest in the El Niño phenomenon (Murphy, 1926).

**1927**: Dutch meteorologist Hendrick Berlage implicated advection of sea surface temperatures across the equatorial Pacific from South America in the mechanism of the Southern Oscillation (Berlage, 1927).

**1928**: Sir Gilbert Walker published an article correlating famines in drought-prone Northeast Brazil with the Southern Oscillation.

**1931**: A paper by Gerhart Schott in 1931 generated interest in El Niño. Schott, like Murphy, wrote about the 1925 El Niño (Schott, 1931).

**1940s–1960s**: There was a marked reduction in Southern Oscillation research, in part because of the lack of a known physical mechanism to explain the phenomenon.

**Mid-1940s–1950s (early years)**: The California sardine fishery collapsed, the impacts of which were represented in John Steinbeck's novels *Cannery Row* and *Sweet Thursday* (Steinbeck, 1945, 1954). California's idled fishing vessels and fish-processing factories were sold to Peruvians and South Africans.

**1953**: A fishmeal factory was shipped to Peru from the collapsed sardine fishery in California. This was Peru's first fishmeal factory.

**1955**: Nylon fishing nets were introduced into the Peruvian anchoveta fishery. Nylon nets did not require the constant repair that the cotton fiber nets needed. This technological development made it easier for entrepreneurs to invest in the fishing sector because fishermen could catch more fish at lower operating costs.

**1956**: Irving Schell linked the Southern Oscillation with variations in sea surface temperature along the Pacific coast of South America, and also linked these variations with upwelling and the thickness of the mixed (top) layer of the ocean (Schell, 1956).

**1957**: Berlage produced a comprehensive description of the Southern Oscillation, and linked it to El Niño, citing the work of both Murphy and Schott (Berlage, 1957).

**1957–58**: Year of the IGY (International Geophysical Year). Data collected incidentally in this international year of scientific research (really a year and a half) later proved to be instrumental for making advances in El Niño research.

**1959**: A CalCOFI (California Cooperative Oceanic Fisheries Investigations) conference was held at Rancho Santa Fe, California, for which a report was prepared on the impact and significance of the 1957–58 El Niño event. This was an important event in generating broader interest in El Niño because it recognized then that El Niño had effects along the California coast and was not just a local phenomenon with local effects off the coast of Peru (Wooster, 1959).

**1950s (late), 1960s (early)**: Bjerknes was contracted by the International Tuna Commission to assess the impacts of El Niño on Pacific tuna fisheries.

**1960**: The Peruvian government and the UN Food and Agriculture Organization (FAO) created an institute to study marine resources, IREMAR (Instituto de los Recursos Marinos).

**1961**: El Niño is mentioned for the first time in the USA in a brief *New York Times* news article about a science paper presented by H. Berlage at a US science conference.

**1964**: In July IMARPE (Instituto del Mar del Peru) was created in Callao, Peru to replace IREMAR. IMARPE was given the responsibility for planning, directing, executing and coordinating research on marine resources at the national level.

**1965**: The first El Niño event to capture the attention of the Peruvian fishing sector occurred. This sector took notice of El Niño's adverse impacts on the anchoveta population following more than a decade of constant increases by Peruvian fishing boats of anchoveta catches.

**1965**: Australian meteorologist Sandy Troup produced a modern, comprehensive description of the Southern Oscillation (Troup, 1965).

**1965**: An American fisheries scientist suggested that the destruction of Peru's guano bird population (numbered in the multimillions at the time) would free up an extra 2 million tonnes of anchoveta for the fishmeal factories in order to process fishmeal, a highly valuable foreign-exchange-earning export.

**1966+1969**: Bjerknes, using IGY data obtained in 1957–58, explained the linkages of the Southern Oscillation with sea surface temperature changes in the eastern equatorial Pacific, showing that El Niño was not just a local phenomenon but was basin-wide. This was a most important milestone in El Niño research (Bjerknes, 1966, 1969).

**1969**: The International Decade for Ocean Exploration (IDOE) was established for the 1970s.

**1971**: An international Coastal Upwelling Ecosystems Analysis Program (CUEA) was developed. CUEA research focused on several coastal upwelling ecosystems including that of Peru. However, at the outset of the Peruvian coastal upwelling research activity, no reference had been made to El Niño, a major disruptor to coastal upwelling in the eastern equatorial Pacific.

**1972**: Meteorologist Peter Rowntree used a climate model to substantiate Bjerknes' idea that a warmer-than-normal eastern equatorial Pacific sea surface temperatures could cause climate teleconnections in the mid-latitudes (Rowntree, 1972).

**1972–73**: The combination of El Niño, fish population dynamics, and the overfishing of Peru's anchoveta population led to the collapse of the anchoveta population and of Peru's fishing industry. It led to worldwide notice that El Niño's adverse impacts on fishmeal production could affect various countries as the broiler industry in the USA shifted to soymeal, an alternative animal feed supplement. Wheat farmers shifted to soybean production, worsening the global food crisis.

**1973**: The collapse of the fishery led Peru's socialist military government to nationalize the fishing sector.

**1974**: A workshop on "The Phenomenon Known as El Niño," the first major international scientific workshop on El Niño, was organized in Guayaquil, Ecuador, 4–12 December.

**1974**: In late 1974 William Quinn issued a forecast of an El Niño. An event began to develop in early 1975 but then collapsed (Wyrtki *et al.*, 1976). This was the same year that Klaus Wyrtki published a paper about El Niño as an oceanic response to the collapse of the trade winds.

**1974**: A sea level monitoring network (the "brainchild" of Klaus Wyrtki: see Wyrtki, 1979, p. 25) in the equatorial Pacific was developed as a tool to monitor the variability of equatorial currents, but served well to monitor El Niño. The first stations were installed in 1974. This was an important step in El Niño research, because Wyrtki figured out how the equatorial atmosphere and the ocean influence each other.

**1975**: The ERFEN (Estudio Regional del Fenomino El Niño) was created by the governments of Peru, Chile, Ecuador, and Colombia. The goal of ERFEN, as its acronym suggests), is to study all aspects of El Niño and its impacts on the environment and on society.

**1975**: Oceanographer Stuart Godfrey demonstrated that ocean dynamics could explain the warming of the eastern equatorial Pacific when the equatorial easterly winds weakened in the central and western Pacific.

**1976**: Following a few years of poor fishing, the Peruvian government de-nationalized the fishing sector.

**1976**: NOAA's Pacific Marine Environmental Laboratories deployed the first successful moored equatorial current meter buoy.

**1976**: Kevin Trenberth defined the South Pacific Convergence Zone (SPCZ) and its role in ENSO. He also defined the 2–7 year time scale for El Niño (Trenberth, 1976).

**1976–77**: The Peruvian coastal upwelling project (under CUEA) was organized off the coast of Peru. An El Niño developed in 1976.

**1978**: Using proxy (indirect) information, Quinn and his colleagues sought to identify several centuries of El Niño events and their "strengths" (Quinn *et al.*, 1978).

**1979**: Wyrtki expanded the island sea level gauge network throughout the tropical Pacific. This proved essential for observing the dynamics of the ocean during an El Niño episode.

**1979**: NOAA began the EPOCS program (Eastern Pacific Ocean Climate Studies), specifically focused on ENSO. EPOCS, according to NOAA, "pioneered the use of moored buoys that protrude above the sea surface, which yielded the first continuous, long-term observations of surface currents, temperatures, and surface winds in the region".

**1970s (late)**: A massive computer-readable compilation of global marine observations, some dating back to the nineteenth century, was compiled (COADS, Comprehensive Ocean–Atmosphere Data Set).

COADS proved essential in empirical studies of ocean–atmosphere changes in the Pacific during El Niño events.

**1980**: Oceanographer Adrian Gill proposed a theoretical explanation for the enhanced heating of the tropical Pacific atmosphere caused by the increases in sea surface temperatures (Gill, 1980).

**1981**: Atmospheric scientists Peter Webster (1981) and Brian Hoskins and David Karoly (1981) developed theoretical explanations of the teleconnections from the equatorial Pacific sea surface temperatures to the mid-latitude atmospheric circulation.

**1982**: Planning for the development of an international monitoring program TOGA (Tropical Ocean–Global Atmosphere) was just getting under way.

**1982 (March)**: Eugene Rasmusson and Thomas Carpenter published a paper introducing the notion of a "canonical El Niño". They used COADS to document in detail the typical progression of an El Niño event in the tropical atmosphere and ocean. The term "ENSO" (El Niño–Southern Oscillation) was also introduced for the first time (Rasmusson and Carpenter, 1982).

**1982 (October)**: El Niño researchers met at a workshop at Princeton University, New Jersey. The general feeling was that no El Niño would develop at that time. In reality, a major El Niño was already under way.

**1982–83**: The most intense El Niño in a century (until 1982) took place. The onset of this event was not forecast well in advance. It sparked scientific research and government interest in El Niño in several countries. Until the 1997–98 extraordinary El Niño event, this was referred to as the "El Niño of the century".

**1985**: UNEP created a Working Group on the Socioeconomic Aspects of El Niño. UNEP and NCAR convened a workshop in Lugano, Switzerland (November), on the socioeconomic impacts associated with the 1982–83 El Niño. This was a first attempt to identify in a systematic way some of the worldwide societal impacts of El Niño.

**1985**: The 10-year TOGA program was launched. New observing systems for the ENSO cycle in the equatorial Pacific were set up. For the first time several aspects of El Niño could be monitored effectively in real time.

**1986**: Oceanographers Mark Cane and Stephen Zebiak (Lamont–Doherty Earth Observatory, Columbia University) issued the first *public* forecast of the onset of an El Niño. They were correct but were challenged by some scientists for having "gone public" with an experimental forecast produced by their computer model.

**1986–87**: A moderate El Niño occurred, the first El Niño event to have been successfully forecast.

**1987**: Meteorologists Chester Ropelewski and David Halpert published scientific articles on the worldwide seasonal impacts of El Niño. These impacts were called "teleconnections". Most maps of El Niño impacts now are spinoffs of their original teleconnection maps (Ropelewski and Halpert, 1987).

**1988–89**: A moderate (some say strong) La Niña developed in early spring 1988 and was linked to a major drought in the USA ($40 billion damage) by researchers Kevin Trenberth, Grant Branstator and Phillip Arkin (1988). Cane and Zebiak forecast this event, as did other ENSO modeling groups.

**1990**: The now widely recognized jump in 1976–77 in more frequent El Niño events was first identified by Trenberth, raising questions about the role of global warming in influencing the frequency, intensity, and duration of El Niño events (Trenberth, 1990).

**1990**: The World Climate Impacts Program (under UNEP) and NCAR organized the first workshop on "El Niño and Climate Change" in Bangkok, Thailand.

**1990–95**: Within this period, the unusual behavior of sea surface temperatures in the tropical Pacific led Peruvians and some scientists to believe that there were three weak El Niño events in this period.

**1993**: Australia Bureau of Meteorology Research Centre (BMRC) convened a workshop in Melbourne, Australia, focused on "El Niño and Climate Change".

**1994 (December)**: The 10-year research program called TOGA ended. The comprehensive TOGA monitoring system, whose core is 70 TAO moored buoys, was maintained in the tropical Pacific, reporting in real time, and with research products and analyses displayed on the Internet a day later.

**1995**: Climate Variability and Predictability (CLIVAR), an interdisciplinary international research effort with the World Climate Research Programme, was established.

**1995 (November)**: NOAA convened an International (governmental) Forum on El Niño in Washington, DC. This forum launched the development of an International Research Institute for climate prediction, now known as the IRI (based at Columbia University).

**1996**: Kevin Trenberth and Timothy Hoar published an article suggesting that the changes in Pacific sea surface temperatures signified the longest event on record with a possible return period of 2000 years. Once again, the influence of global warming on ENSO events emerged as an important issue (see Trenberth and Hoar, 1996).

**1997**: In response to forecasts mid-year of a major El Niño forming in the tropical Pacific, scores of websites devoted to El Niño appeared on the Internet.

**1997 (November)**: The UN General Assembly, recognizing El Niño's potential for devastating impacts worldwide, passed a resolution (A/52/200) entitled "International Cooperation to Reduce the Impact of the El Niño Phenomenon". The resolution called upon UN agencies to undertake a retrospective assessment of the physical and societal aspects of the 1997–98 ENSO event.

**1997–98**: One of the two strongest El Niño events in the twentieth century began to develop in February 1997. Damage attributed to this event has been estimated by NOAA at US$34 billion, and by Swiss Reinsurance at over US$90 billion.

**1998**: The media, for the first time, seriously discussed La Niña.

**1998 (May)**: A La Niña event was forecast to begin in late 1998 and to continue throughout the winter of 1998–99, into the spring of 1999.

**1998 (July)**: The first workshop focused solely on La Niña was held in Boulder, CO. It was organized by NCAR, United Nations University, UNEP, and the National Science Foundation, and was carried live on the Internet each day by San Francisco's Exploratorium.

**1998 (November)**: the Government of Ecuador, the UN and the Permanent Commission for the South Pacific (CPPS) convened the First Intergovernmental Meeting of Experts – "The International Seminar on the 1997–98 El Niño" in Guayaquil, Ecuador. This was about 25 years after the first international scientific workshop was convened on the same topic and in the same city.

**1999**: A strong La Niña peaked in January and continued throughout the rest of the year and into the year 2000.

**2000–01**: A focus intensified on whether decadal scale oscillations (and not necessarily global climate change) cause alterations in the frequency or intensity of ENSO extreme events.

# References


AAAS (American Association for the Advancement of Science), 1991. *Malaria and Development in Africa*. Available from AAAS Sub-Saharan Africa Program, 1200 New York Avenue NW, Washington, DC 20005.

Aceituno, P., 1992. El Niño, the Southern Oscillation, and ENSO: Confusing names for a complex ocean–atmosphere interaction. *Bulletin of the American Meteorological Society*, **73**, 483–5.

Acosta, José de, 1588. *Historia Natural y Moral de las Indias*, Sevilla. In *Obras del Padre Jose de Acosta*, Biblioteca de Autores Españoles, Madrid, 1954.

Allen, J. E., 1998. El Niño may be getting warmed up. *Seattle Times*, 2 February.

Ångström, A., 1935. Teleconnections of climate changes in present time. *Geografiska Analer*, **17**, 242–58.

Archibald, E. D., 1900. Droughts, famines, and forecasts in India. *Monthly Weather Review*, June, 246–8.

Arntz, W. E., 1984. El Niño and Peru: positive aspects. *Oceanus*, **27**, 36–9.

Associated Press, 1999. Study asserts El Niño saved lives, had economic benefits. *Washington Post*, 2 September, p. A26.

Bacastow, R. B., J. A. Adams, C. D. Keeling, D. J. Moss, and T. P. Whorf, 1980. Atmospheric carbon dioxide, the Southern Oscillation, and the weak 1975 El Niño. *Science*, **210**, 66–8.

Barnett, T. P., 1977. An attempt to verify some theories of El Niño. *Journal of Physical Oceanography*, **7**, 633–47.

Barnston, A. G., H. M. van den Dool, S. E. Zebiak, T. P. Barnett, Ming Ji, D. R. Rodenhuis, M. A. Cane, A. Leetmaa, N. E. Graham, C. R. Ropelewski, V. E. Kousky, E. A. O'Lenic, and R. E. Livezey, 1994. Long-lead seasonal forecasts – where do we stand? *Bulletin of the American Meteorological Society*, **75**, 2097–114.

Barnston, A. G., M. H. Glantz, and Y. He, 1999. Predictive skill of statistical and dynamical climate models in SST forecasts during the 1997–98 El Niño episode and the 1998 La Niña onset. *Bulletin of the American Meteorological Society*, **80**, 217–43.

Barsugli, J. J., J. S. Whitaker, A. F. Loughe, P. D. Sardeshmukh, and Z. Toth, 1999. The effect of the 1997–98 El Niño on individual large-scale weather events. *Bulletin of the American Meteorological Society*, **80**, 1399–411.

Beck, L. R., M. H. Rodriguez, and S. W. Dister, 1997. Assessment of a remote

sensing-based model for predicting malaria transmission risk in villages of Chiapas, Mexico. *American Journal of Tropical Medicine and Hygiene*, **56**, 99–106.
Bell, G. D. and M. S. Halpert, 1998. Climate assessment for 1997. Supplement, *Bulletin of the American Meteorological Society*, **79**(5), S1–S50.
Berlage, H. P., 1927. East-monsoon forecasting in Java. *Verhandelingen Koninklijk Magnetisch en Meteorologisch Observatorium te Batavia,* **20**.
Berlage, H. P., 1957. Fluctuations of the general atmospheric circulation of more than one year, their nature, and prognostic value. *Royal Netherlands Meteorological Institute Yearbook*, **69**, 151–9.
Bjerknes, J., 1961. El Niño study based on analysis of ocean surface temperatures, 1935 to 1957. *Inter-American Tropical Tuna Commission Bulletin*, **5**, 219–303.
1966. A possible response of the atmosphere Hadley circulation to equatorial anomalies of ocean temperature. *Tellus*, **8**, 820–9.
1969. Atmospheric teleconnections from the equatorial Pacific. *Monthly Weather Review*, **97**, 163–72.
Blanford, H. G., 1884. On the connection of the Himalayan snowfall with dry winds and seasons of drought in India. *Proceedings of the Royal Society of London*, **37**, 3–22.
Blimsrieder, M., 1998. El Niño of 1982–1983 in Galápagos. Isla Santa Cruz, Galápagos: Charles Darwin Research Station. www.darwinfoundation.org or contact cdrs@fcdarwin.org.ec (accessed December 1998).
Bouma, M. J. and C. Dye, 1997. Cycles of malaria associated with El Niño in Venezuela. *Journal of the American Medical Association*, **278**, 1772–4.
Bouma, M. F., H. E. Sondorp, and H. J. van der Kaay, 1994. Climate change and periodic epidemic malaria. *Lancet*, **343**, 1140.
Bouma, M. F., R. S. Kovats, S. A. Goubet, J. Cox, A. Hines, 1997. Global assessment of El Niño's disaster burden. *Lancet*, **350**, 1435–8.
Broad, K., 1999. Climate, culture, and Peruvian fisheries: the El Niño of 1997–98. Doctoral thesis. Lamont–Doherty Earth Observatory, Palisades, NY.
Brooks, H., 1986. The typology of surprises in technology, institutions, and development. In *Sustainable Development of the Biosphere*, ed. W. C. Clark and R. E. Munn, p. 326. Cambridge: Cambridge University Press.
Brown, B. G. and R. W. Katz, 1991. The use of statistical methods in the search for teleconnections: past, present, and future. In *Teleconnections Linking Worldwide Climate Anomalies*, ed. M. H. Glantz, R. W. Katz, and N. Nicholls, pp. 371–400. Cambridge: Cambridge University Press.
Brown, L. and E. P. Eckholm, 1974. *By Bread Alone*. New York: Praeger Press.
Burton, I., R. W. Kates, and G. F. White, 1993. *The Environment as Hazard.* New York: Oxford University Press.
Busalacchi, A., 1998. Monitoring La Niña. In *A La Niña Summit: A Review of the Causes and Consequences of Cold Events*, Workshop Report, ed. Michael Glantz, held 15–17 July 1998 in Boulder, CO (National Center for Atmospheric Research, Environmental and Societal Impacts Group: Boulder, Colorado) p. 32.
CAC (Climate Analysis Center), 1995. *Climate Diagnostics Bulletin*, December

1994. Washington, DC: National Weather Service, NOAA.

CalCOFI (California Cooperative Oceanic Fisheries Investigations), 1959. Editor's summary of the symposium. *Reports*, vol. VII, 1 January 1958 to 30 June 1959, pp. 211–18.

Cane, M. A., 1991. Forecasting El Niño with a geophysical model. In *Teleconnections Linking Worldwide Climate Anomalies*, ed. M. H. Glantz, R. W. Katz, and N. Nicholls, pp. 345–70. Cambridge: Cambridge University Press.

Cane, M. A., S. E. Zebiak, and S. C. Dolan, 1986. Experimental forecasts of El Niño. *Nature*, **321**, 827–32.

Carrillo, C. 1892. Disertación sobre las Corrientes Océanicas y Estudios de la Corriente Peruana de Humboldt. *Boletines del Sociedad Geográfico Lima*, **11**, p. 84. Microfiche.

Caviedes, C. N., 1985. Emergency and institutional crisis in Peru during El Niño 1982–83. *Disasters*, **9**, 70–4.

CBS News, 1997. November polling information. Available on-line website at www.cbsnews.com

CDRS (Charles Darwin Research Station), 1997. Report on the status of infrastructure in visitor sites of Galápagos National Park, September.

Changnon, S. A., 1999. Impacts of 1997–98 El Niño-generated weather in the United States. *Bulletin of the American Meteorology Society*, **80**, 1819–27.

Coffroth, M. A., H. R. Lasker, and J. K. Oliver, 1990. Coral mortality outside of the eastern Pacific during 1982–1983: Relationship to El Niño. In *Global Ecological Consequences of the 1982–83 El Niño–Southern Oscillation*, ed. P. W. Glynn, pp. 141–82. New York: Elsevier.

Coker, R. E., 1908. The fisheries and the guano industry of Peru. *Bulletin of the Bureau of Fisheries*, **28**, Part 1, 333–65.

1920. Peru's wealth-producing birds: vast riches in the guano deposits of cormorants, pelicans, and petrels which nest on her barren, rainless coast. *National Geographic Magazine*, **37**, 536–66.

Colwell, R. R., 1996. Global climate and infectious disease: the cholera paradigm. *Science*, **274**, 2025–31.

Colwell, R. R. and J. A. Patz, 1998. Climate, Infectious Disease and Health: An Interdisciplinary Perspective. Report based on an American Academy of Microbiology Colloquium held 20–22 June 1997, Montero Bay, Jamaica. Washington, DC: American Academy of Microbiology.

Conner, S. J., M. C. Thomson, S. R. Flasse, and R. H. Perryman, 1998. Environmental information systems in malaria risk mapping and epidemic forecasting. *Disasters*, **22**, 39–56.

CORECA (Consejo Regional de Cooperación Agricola), 1997. *Plan para mitigar los efectos del fenómeno de El Niño en el Sector Agropecuario*. San José, Costa Rica: MAG (Ministro de Agricultura y Ganaderis).

Cornejo-Grunauer, P., 1998. The ENSO phenomenon of 1997–98. Paper presented at the IVth Ecuador Congress on Aquaculture, 22–27 October, Guayaquil.

Cunha, E. da, 1944. *Rebellion in the Backlands*. Translated from *Os Sertoes* by Euclides da Cunha. Chicago, IL: University of Chicago Press.

Cushing, D. H., 1982. *Climate and Fisheries*. London: Academic Press.

Diaz, H. F. and F. Markgraf (eds.), 1992. *El Niño: Historical and Paleoclimatic Aspects of the Southern Oscillation*. Cambridge: Cambridge University Press.

*Economist*, 1997. Fujimori against El Niño. 27 September, pp. 35–6.

Eguiguren, D. V., 1894. Las lluvias en Piura [The rains in Piura]. *Boletin del Sociedad Geográfico de Lima,* **4**, 241–58.

1895. Estudios sobre la riqueza territorial de la provincia de Piura. *Boletin del Sociedad Geográfico Lima*, **11**, p. 84. Microfiche.

*El Universo*, 1997. Según congresista peruano: El Niño impedirá conflicto. 26 October, p. 3.

Enfield, D. B., 1989. El Niño, past and present. *Reviews of Geophysics*, **27**(1), 159–87.

ENN (Environmental News Network), 1998. El Niño Special Report, from website www.enn.com/elnino/

Epstein, P. R. (ed.), 1999. *Extreme Weather Events: The Health and Economic Consequences of the 1997/98 El Niño and La Niña*. Cambridge, MA: Center for Health and the Global Environment, Harvard Medical School.

Epstein, P. R., O. C. Pena, and J. B. Racedo, 1995. Climate and disease in Colombia. *Lancet*, **346**, 1243–4.

Epstein, P. R., A. Haines, P. Reiter, 1998. Global warming and vector-borne disease. *Lancet*, **351**, 1737–8.

Fagan, B., 1999. *Floods, Famines and Emperors: El Niño and the Fate of Civilization*. New York: Basic Books.

Feldman, G., 1985. Satellites, seabirds, and seals. In *El Niño en las Galápagos: El Evento de 1982–83*, ed. G. Robinson and E. M. del Pino, pp. 125–30. Quito: Fundación Charles Darwin para las Islas Galápagos.

Flannery, T. F., 1995. *The Future Eaters: An Ecological History of the Australasian Island and People*. New York: George Braziller, Inc.

Flohn, H. and H. Fleer, 1975. Climatic teleconnections with the equatorial Pacific and the role of ocean/atmosphere coupling. *Atmosphere*, **13**, 96–109.

Flores-Palomino, M., 1998. Climatic variability because of El Niño phenomenon and its incidence in Peruvian fishing sector. In *An Assessment of the Use of Remote Sensing and Other Information Related to ENSO: The Use of ENSO Information in Peru*. NASA/NCAR/Peru Project Final Report. Boulder, CO: Environmental and Societal Impacts Group, NCAR.

Freeman, T. and M. Bradley, 1996. Temperature is predictive of severe malaria years in Zimbabwe. *Transactions of the Royal Society of Tropical Medicine and Hygiene*, **90**, 232.

FUNCEME (Fundação Cearense de Meteorologia e Recursos), 1992. *Monitor Climatico*, **6**. Fortaleza, Brazil: FUNCEME.

Garcia, R., 1981. *Drought and Man, the 1972 Case History*, vol. 1. *Nature Pleads Not Guilty*. New York: Pergamon Press.

Garnham, P. C., 1948. The incidence of malaria at high altitudes. *Journal of the National Malaria Society*, **7**, 275–84.

Ghil, M. and S. Childress, 1987. *Topics in Geophysical Fluid Dynamics: Atmospheric Dynamics*. Berlin: Springer-Verlag.

Gill, A. E., 1980. Some simple solutions for heat-induced tropical circulation.

*Quarterly Journal of the Royal Meteorological Society,* **106**, 447–62.

Gill, A. E. and E. M. Rasmusson, 1983. The 1982–83 climate anomaly in the equatorial Pacific. *Nature*, **306**, 229–34.

Glantz, M. H., 1982. Consequences and responsibilities in drought forecasting: the case of Yakima, 1977. *Water Resources Research*, **18**, 3–13.

1998a. *An Assessment of the Use of Remote Sensing and Other Information Related to ENSO: The Use of ENSO Information in Peru.* NASA/NCAR/Peru Project Final Report. Boulder, CO: Environmental and Societal Impacts Group, NCAR.

1998b. *A Review of the Causes and Consequences of Cold Events: A La Niña Summit.* Executive Summary of the Workshop held 15–17 July 1998 in Boulder, CO. Boulder: National Center for Atmospheric Research. Website www.dir.ucar.edu/esig/lanina/

Glantz, M. H. and J. D. Thompson, 1981. *Resource Management and Environmental Uncertainty: Lessons from Coastal Upwelling Fisheries.* New York: John Wiley & Sons.

Glynn, P. W. (ed.), 1990. *Global Ecological Consequences of the 1982–83 El Niño Southern Oscillation.* New York: Elsevier.

Gray, W. M., 1993. *Forecast of Atlantic Seasonal Hurricane Activity for 1993.* Fort Collins: Department of Atmospheric Sciences, Colorado State University.

Greenpeace International, 1994. *The Climate Time Bomb: Signs of Climate Change from the Greenpeace Database.* Amsterdam: Stichting Greenpeace Council.

Gueri, M., C. Gonzalez, and V. Moran, 1986. The effect of the floods caused by "El Niño" on health. *Disasters*, **10**, 118–24.

Halpert, M. S., G. D. Bell, V. E. Kousky, and C. F. Ropelewski 1994. *Fifth Annual Climate Assessment 1993.* Camp Springs, MD: Climate Analysis Center, National Weather Service.

Hammer, G., 1995. ENSO impacts in Australia. *The ENSO Signal*, **3**, p. 5. Silver Spring, MD: NOAA/OGP.

Hammergren, L. A., 1981. Peruvian political and administrative responses to El Niño: organizational, ideological and political constraints. In *Resource Management and Environmental Uncertainty: Lessons from Coastal Upwelling Fisheries*, ed. M. H. Glantz and J. D. Thompson, pp. 317–50. New York: John Wiley & Sons.

Handler, P. and E. Handler, 1983. Climatic anomalies in the tropical Pacific Ocean and corn yields in the United States. *Science*, **220**, 1155–6.

Hansen, J. E., 1988. In article by Eugene Linden, Big chill for the greenhouse: remember El Niño? Now comes its cool sibling, La Niña. *Time Magazine*, 31 October, p. 90.

1990. Physical aspects of the El Niño event of 1982–83. In *Global Ecological Consequences of the 1982–83 El Niño Southern Oscillation*, ed. P. W. Glynn, pp. 1–19. New York: Elsevier.

Hare, S. R., 1998. Recent El Niño brought downpour of media coverage. *EOS, Transactions of the AGU*, **79**, 481.

Harrison, D. E. and M. A. Cane, 1984. Changes in the Pacific during the 1982–83 event. *Oceanus*, **27**(2), 21–28.

Hildebrandsson, H. H., 1897. East-Monsoon forecasting in Java. *Verhandelingen Koninklijk Magnetisch en Meteorologisch Observatorium to Bataviam* **26**.

Holling, C. S., 1986. The resilience of terrestrial ecosystems: local surprise and global change. In *Sustainable Development of the Biosphere*, ed. W. C. Clark and R. E. Munn, pp. 292–317. Cambridge: Cambridge University Press.

Hoskins, B. J., and D. J. Karoly, 1981. Steady linear response of a spherical atmosphere to thermal and orographic forcing. *Journal of the Atmospheric Sciences*, **38**, 1179–96.

Houghton, J. T., G. L. Jenkins, and J. J. Ephraums (eds.), 1990. *Climate Change: The IPCC Scientific Assessment*. Cambridge: Cambridge University Press.

Houghton, J. T., L. G. Meira Filho, B. A. Callander. N. Harris, A. Kattenberg, and K. Maskell (eds.), 1996. *Climate Change 1995: The Science of Climate Change*. Cambridge: Cambridge University Press.

IDNDR (International Decade for Natural Disaster Reduction), 2000. *Program Forum 1999*. Proceedings of conference held in Geneva, 5–9 July 1999. Geneva, Switzerland: IDNDR Secretariat, in press. Website: www.idndr.org/forum/

IPCC (Intergovernmental Panel on Climate Change), 1990. *Climate Change: The IPCC Scientific Assessment*, ed. J. T. Houghton, G. J. Jenkins, and J. Ephraums. Cambridge: Cambridge University Press.

1996. *Climate Change 1995: The Science of Climate Change*, ed. J. T. Houghton, L. G. Meira Filho, B. A. Callander, N. Harris, A. Kattenberg, and K. Maskell. Contribution of Working Group II to Second Assessment Report. Cambridge: Cambridge University Press.

Jacobs, G. A., H. E. Hurlburt, H. C. Kindle, E. J. Metzger, J. L. Mitchell, W. J. Teague, and A. J. Wallcraft, 1994. Decade-scale trans-Pacific propagation and warming effects of an El Niño anomaly. *Nature*, **370**, 360–3.

Jordán, R., 1991. Impact of ENSO events on the southeastern Pacific region with special reference to the interaction of fishing and climate variability. In *Teleconnections Linking Worldwide Climate Anomalies*, ed. M. H. Glantz, R. W. Katz, and N. Nicholls, pp. 401–30. Cambridge: Cambridge University Press.

Kahya, E. and J. A. Dracup, 1993. U.S. streamflow patterns in relation to the El Niño/Southern Oscillation. *Water Resources Research*, **29**, 2491–503.

Kerr, R. A., 1982. U.S. weather and the equatorial connection. *Science*, **216**, 609.

1994. Official forecasts pushed out to a year ahead. *Science*, **266**, 1940–1.

1998. Models win big in forecasting El Niño. *Science*, **280**, 522–3.

Kessler, W. S. and M. J. McPhaden, 1995. Oceanic equatorial waves and the 1991–93 El Niño. *Journal of Climate*, **8**, 1757–74.

Kestin, T. and N. Nicholls, 1998. Forecasting with ENSO in Australia: the problems are not over yet! In *A La Niña Summit: Review of the Causes and Consequences of Cold Events*, ed. M. Glantz. Report of the Workshop held 15–17 July 1998 in Boulder, CO, pp. 80–1. Boulder, CO: Environmental and Societal Impacts Group, National Center for Atmospheric Research.

Kiladis, G. N. and H. van Loon, 1988. The Southern Oscillation. Part VII: Meteorological anomalies over the Indian and Pacific sectors associated with extremes of the Oscillation. *Monthly Weather Review*, **116**, 120–36.

Koblinsky, C. J., P. Gaspar, and G. Lagerloef (eds.), 1992. *The Future of Spaceborne Altimetry: Oceans and Climate Change*. Washington, DC: Joint Oceanographic Institutions, Inc.

Kovats, R. S., M. J. Bouma, and A. Haines, 1999. *El Niño and Health*. WHO/SDE/PHE/99.4. Geneva: World Health Organization.

Lagos, P. and J. Buizer, 1992. El Niño and Peru: a nation's response to interannual climate variability. In *Natural and Technological Disasters: Causes, Effects and Preventive Measures*, ed. S. K. Mujumdar, G. S. Forbes, E. W. Miller, and R. F. Schmalz, pp. 223–8. Philadelphia: Pennsylvania Academy of Sciences.

Lau, K.-M. and A. J. Busalacchi, 1993. El Niño-Southern Oscillation: A view from space. In *Atlas of Satellite Observations Related to Global Change*, ed. R. J. Gurney, J. L. Foster, and C. L. Parkinson, pp. 281–96. Cambridge: Cambridge University Press.

Lemonick, M., 1997. Writing about astronomy. In *A Field Guide for Science Writers*, ed. D. Blum and M. Knudson, pp. 196–202. New York: Oxford University Press.

Lien, T. V. and N. H. Ninh, 1996. Dengue fever and ENSO events in Vietnam. Hanoi, Vietnam: CERED (Center for Environment Research, Education and Development). Mimeo.

Linden, E., 1988. Big chill for the Greenhouse. *Time Magazine*, 31 October, p. 90.

Lockyer, N., and J. S. Lockyer, 1904. The behavior of the short-period atmospheric pressure variation over the Earth's surface. *Proceedings of the Royal Society of London*, **73**, 457–69.

Loevinsohn, M., 1994. Climatic warming and increased malaria incidence in Rwanda. *Lancet*, **343**, 714–18.

McKay, G. and T. Allsopp, 1976. Global interdependence of the climate of 1972. In *Proceedings of the Mexican Geophysical Union Symposium on Living with Climate Change*, Mexico City, May, pp. 79–86.

McMichael, A. J., A. Haines, R. Slooff, and S. Kovats (eds.), 1996. *Climate Change and Human Health*. Geneva: World Health Organization.

McPhaden, M. J., 1999. Genesis and evolution of the 1997–98 El Niño. *Science*, **283**, 950–54.

McPhaden, M. J., A. Busalacchi, R. Cheney, J. Donguy, K. Gage, D. Halpern, M. Ji, P. Julian, G. Meyers, G. Mitchum, P. Niiler, Jo Picaut, R. Reynolds, N. Smith, and K. Takeuchi, 1998. The Tropical Ocean-Global Atmosphere observing system: a decade of progress. *Journal of Geophysical Research*, **103**(C7), 14169–240.

*Mosaic*, 1975. All that unplowed sea. A *Mosaic* Special Issue: Food. *Mosaic*, **6**(3), 22–7.

Murphy, R. C., 1923. The oceanography of the Peruvian littoral with reference to the abundance and distribution of marine life. *Geographical Review*, **13**, 64–85.

1926. Oceanic and climatic phenomena along the west coast of South America during 1925. *Geographical Review*, **16**, 26–53.

1954. The guano and the anchoveta fishery. Reprinted in *Resource Management and Environmental Uncertainty: Lessons from Coastal Upwelling Fisheries*, ed. M. H. Glantz and J. D. Thompson, pp. 81–106. New York: John Wiley &

Sons, 1981.

Myers, N., 1995. Environmental unknowns. *Science*, **269**, 358–60.

NASA/JPL, 1997. Independent NASA satellite measurements confirm El Niño is back and strong. Press release 16 September 1997, received by Internet from news-release@www-onlab.jpl.nasa.gov

Nicholls, N., 1986. A method for predicting Murray Valley encephalitis in southeast Australia using the Southern Oscillation. *Australian Journal of Experimental Biological and Medical Science*, **64**, 587–94.

1987. The El Niño/Southern Oscillation phenomenon. In *Climate Crisis*, ed. M. H. Glantz, R. W. Katz, and M. E. Krenz, pp. 2–10. New York: United Nations Publications.

1988. El Niño–Southern Oscillation impact prediction. *Bulletin of the American Meteorological Society*, **69**, 173–6.

1993. ENSO, drought and flooding rain in South-East Asia. In *South-East Asia's Environmental Future: The Search for Sustainability*, ed. H. Brookfield and Y. Byron, pp. 154–75. Tokyo, Japan: United Nations University Press and Oxford University Press.

Nicholls, N. and T. Kestin, 1998. Communicating climate. An editorial comment. *Climatic Change*, **40**, 417–20.

NMSA (National Meteorological Services Agency, Ethiopia), 1987. *The Impact of El Niño on Ethiopian Weather*, Report, December.

NOAA (National Oceanic and Atmospheric Administration), 1997a. El Niño–Southern Oscillation ENSO Advisory, 13 March, 97/2. Washington, DC: Climate Prediction Center.

1997b. *El Niño Southern Oscillation ENSO Advisory*, 26 June, No. 97/5. Washington, DC: Climate Prediction Center.

NRC (National Research Council), 1990. *TOGA: A Review of Progress and Future Opportunities*, pp. 11–12. Washington, DC: National Academy Press.

NSF (National Science Foundation), 1998. El Niño and climate more predictable than previously thought. NSF Press Release 98–69. Arlington, VA: NSF/ Office of Legislative and Public Affairs.

Orlove, B. S. and J. L. Tosteson, 1999. The application of seasonal-to-interannual climate forecasts based on El Niño-Southern Oscillation (ENSO) events: lessons from Australia, Brazil, Ethiopia, Peru, and Zimbabwe. Proceeds of a Workshop on Environmental Politics. Berkeley, CA: Institute of International Studies, University of California, Berkeley.

*Oxford English Dictionary*, 1971. *The Compact Edition of the Oxford English Dictionary.* Complete text reproduced micrographically. Volume II, P–Z. Oxford, UK: Oxford University Press.

Parry, M. L., T. R. Carter, and N. T. Konjin (eds.), 1988. *The Impact of Climatic Variations on Agriculture*, vol. 2 *Assessments in Semi-Arid Regions*. Dordrecht: Riedel Publishers.

Partridge, I. J., 1991. *Will It Rain?* Brisbane: Department of Primary Industries.

Patz, J. A., Martens, W. J.M. , D. A. Fochs, and T. H. Jetten, 1998. Dengue fever epidemic potential as projected by general circulation models of global climate change. *Environmental Health Perspectives*, **106**, 147–53.

Paulik, G. J., 1981. Anchovies, birds, and fishermen in the Peru current. In *Resource Management and Environmental Uncertainty: Lessons from Coastal Upwelling Fisheries*, ed. M. H. Glantz and J. D. Thompson, pp. 35–79. New York: John Wiley & Sons.

Petterssen, S., 1969. *Introduction to Meteorology*, 3rd edn. New York: McGraw Hill.

Pezet, F. A., 1895. The counter-current "El-Niño," on the coast of northern Peru. *Boletines del Sociedad Geográfico Lima*, **11**, 603–6.

Pfaff, A., K. Broad, and M. Glantz, 1999. Who benefits from climate forecasts? *Nature*, **397**, 645–6.

Philander, S. G., 1990. *El Niño, La Niña, and the Southern Oscillation*. San Diego, CA: Academy Press.

1998. Who is El Nino? *Eos*, **79**(13), 170.

Pielke, R. A., Jr and C. W. Landsea, 1999. La Niña, El Niño, and Atlantic hurricane damages in the United States. *Bulletin of the American Meteorological Society*, **80**, 2027–33.

Preston, R., 1995. *The Hot Zone*. New York: Anchor Books Doubleday.

*Quarterly Journal of the Royal Meteorological Society*, 1959. Quote from obituary of Sir Gilbert Walker. **85**, 186.

Quinn, W. H., D. O. Zopf, K. S. Short, and R. T. Kuo Yang, 1978. Historical trends and statistics of the Southern Oscillation, El Niño, and Indonesian droughts. *Fisheries Bulletin*, **76**, 663–78.

Quinn, W., V. T. Neal, and S. E. A. Mayolo, 1987. El Niño occurrences over the past four and a half centuries. *Journal of Geophysical Research*, **92**, C13, 14449–61.

Rasmusson, E. M., 1984a. El Niño: The ocean/atmosphere connection. *Oceanus*, **27**(2), 5–13.

1984b. Meteorological aspects of El Niño–Southern Oscillation. In *Proceedings of the 15th Conference on Hurricanes and Tropical Meteorology*, pp. 17–20. Boston, MA: American Meteorological Society.

Rasmusson, E. M. and P. A. Arkin, 1985. Interannual climate variability associted with the El Niño/Southern Oscillation. In *Coupled Ocean–Atmosphere Models*, ed. J. C. J. Nihoul, Elsevier Oceanographic Series, vol. 40, pp. 697–725. New York: Elsevier.

Rasmusson, E. M. and T. H. Carpenter, 1982. Variations in tropical sea surface temperature and surface wind fields associated with the Southern Oscillation/El Niño. *Monthly Weather Review*, **110**, 354–84.

Rasmusson, E. M. and J. M. Wallace, 1983. Meteorological aspects of the El Niño/Southern Oscillation. *Science*, **222**, 1195–202.

Rensberger, B., 1997. Covering science for newspapers. In *A Field Guide for Science Writers*, ed. D. Blum and M. Knudson, pp. 8–9. New York: Oxford University Press.

Ropelewski, C. F., 1992. Predicting El Niño events. *Nature*, **356**, 476–7.

Ropelewski, C. F. and M. S. Halpert, 1987. Global and regional scale precipitation patterns associated with the El Niño/Southern Oscillation. *Monthly Weather Review*, **115**, 1606–26.

Rowntree, P. R., 1972. The influence of tropical east Pacific Ocean temperatures on the atmosphere. *Quarterly Journal of the Royal Meteorological Society*, **98**, 290–321.

Sagan, C., 1997. Foreword. In *A Field Guide for Science Writers*, ed. D. Blum and M. Knudson, p. vii. New York: Oxford University Press.

*Samudra Report*, 1999. Waiting for El Niño. **22** (April), 20–23. Also on-line at www.icsf.net/icsf/english/issue_22/art05.htm

Schell, I. I., 1956. The origin and possible prediction of the fluctuations in the Peru Current and upwelling. *Journal of Geophysical Research*, **70**, 5529–40.

Schott, G., 1931. Peru Strom und Seine nordlichen Nachbargebiete im normaler und abnormaler Ausbildung. *Annales Hydrographiques*, **59**, 161–19 [in German].

Sears, A. F., 1895. The coastal desert of Peru. *Bulletin of American Geographical Society*, **28**, 256–71.

Serra, R., 1987. Impact of the 1982–83 ENSO on the Southeastern Pacific fisheries, with emphasis on Chilean fisheries. In *The Societal Impacts Associated with the 1982–83 Worldwide Climate Anomalies*, ed. M. Glantz, R. W. Katz and M. Krenz, pp. 24–9. Boulder, CO: National Center for Atmospheric Research.

Sharma, C., 1999. Waiting for El Niño! *Samudra*, 22 April, 20–3.

Solomon, N., 1998. El Bunko. Website: www.labridge.com/change-links/elbunko.html

Sponberg, K., 1999. Weathering a storm of global statistics. *Nature*, **400**, 13.

Steinbeck, J., 1945. *Cannery Row*. New York: Viking Press.

1954. *Sweet Thursday*. New York: Viking Press.

Stewart, T. R., 1997. Judgment and Decision Research and Climate Surprises. Mimeo. Contribution to an NCAR/Argonne National Labs project for DOE on Climate Surprise. Draft. 27 June, p. 4.

Stolz, W. and M. Sanchez, 1998. Costa Rica: El Niño, the press and public weather services. *World Meteorological Organization Bulletin*, **47**(3), 256–8.

Tarr, R. S. and F. M. McMurry, 1904. *A Complete Geography*. London: The MacMillan Co.

Taylor, P. and S. L. Mutambu, 1986. A review of the malaria situation in Zimbabwe with special reference to the period 1972–1981. *Transactions of the Royal Society of Tropical Medicine and Hygiene*, **80**, 12–19.

Telleria, A. V., 1986. Health consequences of floods in Bolivia in 1982. *Disasters*, **10**, 88–106.

Thompson, D. F., J. B. Malone, and M. Harb, 1996. Bancroftian filariasis distribution in the southern Nile delta: correlation with diurnal temperature differences from satellite imagery. *Emerging Infections Diseases*, **3**, 234–5.

Thompson, J. D., 1977. Ocean deserts and ocean oases. In *Desertification: Environmental Degradation in and around Arid Lands*, ed. M. H. Glantz, pp. 103–39. Boulder, CO: Westview Press.

Thompson, M., R. Ellis, and A. Wildavsky, 1990. *Cultural Theory*. Boulder, CO: Westview Press.

Timmermans, D., J. Kievit, and H. van Bockel, 1996. How do surgeons' probability estimates of operative mortality compare with a decision analytic model? *Acta*

*Psychologica*, **93**, 107–20.

Todd, C., 1893. Meteorological work in Australia: a review. *Australasian Association for the Advancement of Science.* Report of meeting, 5, Adelaide, pp. 246–70.

Tomczak, M. and J. S. Godfrey, 1994. *Regional Oceanography: An Introduction.* Oxford, UK: Pergamon Press.

Trager, J., 1975. *The Great Grain Robbery*. New York: Ballantine Books.

Trenberth, K. E., 1976. Spatial and temporal variations of the Southern Oscillation. *Quarterly Journal of the Royal Meteorological Society*, **102**, 639–53.

1988. Quoted in article written by Eugene Linden, Big chill for the greenhouse: Remember El Niño? Now comes its cool sibling, La Niña. *Time Magazine*, 31 October, p. 90.

1990. Recent observed interdecadal climate changes in the northern hemisphere. *Bulletin of the American Meteorological Society*, **71**, 988–93.

1997. The definition of El Niño. *Bulletin of the American Meteorological Society*, **78**, 12, 2772–7.

1999. The extreme weather events of 1997 and 1998. *Consequences*, **5**(1), 3–15.

2000. Quote in Chapter 7 of the IPCC (Intergovernmental Panel on Climate Change) 2000 Assessment. Cambridge, UK: Cambridge University Press. (In press).

Trenberth, K. and T. J. Hoar, 1996. The 1990–1995 El Niño–Southern Oscillation event: longest on record. *Geophysical Research Letters*, **23**, 57–60.

Trenberth, K. E., G. W. Branstator, and P. A. Arkin, 1988. Origins of the 1988 North American drought. *Science*, **242**, 1640–5.

Tribbia, J., 1995. What the Southern Oscillation is: an atmospheric perspective. In *Usable Science II: The Potential Use and Misuse of El Niño Information in North America*, ed. M. H. Glantz, pp. 18–19. Proceedings of a Workshop held 31 October–3 November 1994 in Boulder, CO. Boulder, CO: National Center for Atmospheric Research.

Troup, A. J., 1965. The Southern Oscillation. *Quarterly Journal of the Royal Meteorological Society*, **91**, 490–506.

UN ECLA (United Nations Economic Commission for Latin America), 1984. *The Natural Disasters of 1982–83 in Bolivia, Ecuador and Peru*, Report E/CEPAL/G.1274, 26 January, p. 24).

Walker, G. T., 1924. Correlation in seasonal variations of weather, IX: A further study of world weather. *Memoirs of the Indian Meteorological Department XIV* (Part IX), 275–332.

1936. Seasonal weather and its prediction. *Smithsonian Institute Annual Report 1935*, pp. 117–38.

Washino, R. K. and B. L. Wood, 1994. Application of remote sensing to arthropod vector surveillance and control, *American Journal of Tropical Medicine and Hygiene*, **50** suppl: 134–44.

Webster, P. J., 1981. Mechanisms determining the atmospheric response to sea surface temperature anomalies. *Journal of the Atmospheric Sciences*, **38**, 554–71.

Weiner, J., 1994. *The Beak of the Finch: A Story of Evolution in our Time*. New

York: Knopf.
WMO, (World Meteorological Organization) 1984. The Global Climate System: A critical review of the climate system during 1982–84. Geneva: WMO.
1991. Website: www.wmo.ch/web/wcrp/wcrp-home.
1999. *The 1997–98 El Niño Event: A Scientific and Technical Retrospective*. WMO-No. 905. Geneva, Switzerland: WMO.
Wooster, W. S., 1959. El Niño. *California Cooperative Oceanic Fisheries Investigations, Reports*, vol. VII, 1 January 1958 to 30 June 1959.
Wooster, W. and O. Guillen, 1974. Characteristics of the El Niño in 1972. *Journal of Marine Research*, **32**, 387–404.
World Health Organization (WHO), 1996. *World Health Report 1996: Fighting Disease Fostering Development*. Geneva: WHO.
World Resources Institute, 1998. Website: www.wri.org/wri-pubs
Wyrtki, K., 1979. Sea level variations: monitoring the breath of the Pacific. *Eos*, **60**(3), 25–27.
Wyrtki, K., E. Stroup, W. Patzert, R. Williams, and W. Quinn, 1976. Predicting and observing El Niño. *Science*, **191**, 343–6.
Zapata-Velasco, A. and J. C. Sueiro, 1999. *Naturaleza y polítca: El gobierno y el Fenómeno del Niño en el Perú 1997–98*. IEP series, Colección Mínima 38. Lima: Instituto de Estudios Peruanos.

# Index

Acosta, José de, 90
Africa, 178, 183–7
droughts in, 1, 40, 75, 94, 99, 117
sub-Saharan, 2–3
analogies, use of, 51–2
anchoveta, 5, 33–4, 38
fishery collapse, 36–7, 39, 167–8
Andrew, Hurricane, 1, 51
Ångström, Anders, 133
Archibald, Douglas, 230
Aristotle, 66
Arkin, Phillip, 85, 235
Arntz, Wolf, 92
Australia, x, 16, 22, 42–6, 54, 62, 75, 96, 110–11, 116–17
drought in, 3–4, 40, 66, 77, 99, 102–3, 134–5
dust storms in, 87
media interest in, 193, 195, 200–1

Barnston, Tony, 124
Bell, Gerald, 93
Berlage, Hendrick, 230
biometeorology, 181
Bjerknes, Jacob, 29, 59, 103, 210, 231–2
Blanford, Henry, 42, 229
Bouma, M. F., 223
Branstator, Grant, 235
Brazil, 97, 191
droughts in, 2–3, 48, 83, 94, 109–10, 134
northeast, 109–10, 134
Brooks, Harvey, 203
Bureau of Meteorology Seasonal Outlooks, 116–17
Busalacchi, Antonio, 169–70

California Sardine Fishery, 36
Callao, 38, 46, 90
Cane, Mark, 85, 105, 171, 205, 234
Cane–Zebiak model, 108–9, 205
canonical El Niño, 84
definition of, 54–6
Carpenter, Thomas, 234
Carrillo, Camilo, 15, 165
CBS News, 193
Ceará, 109–10
Center for Ocean, Land and Atmosphere, 125
Chad, Lake, 134
Changnon, Stanley, 68
chapter overview, 9–10
Charles Darwin Research Station, 146, 162
Chincha Islands, 30–1
cholera, 178, 182–4
chronology of interest in El Niño, 229–36
climate
change, *see* global warming
change and El Niño, 144–5
outlook forums, 135
maps of, 136–7
Climate Prediction Center, 124, 164
CLIVAR (climate variability), 235–6
coastal
ecosystems analysis program (CUEA), 232
populations, 23
upwelling, *see* upwelling
Coker, R. E., 230
Columbia University, 119, 235
Comprehensive Ocean–Atmosphere Data Set, 233
Conference of Parties, Kyoto, 193
coupled models, 106, 172
Cunha, Euclides da, 109
Cushing, David, 33

definition of El Niño, 19
dengue fever, 178, 183, 185–7
diagram
of El Niño conditions, 70
of La Niña conditions, 71
of normal conditions, 70
of sea level changes 1987–88, 64
disease and satellite images, 186

drought
in Africa, *see* Africa
in Australia, *see* Australia
in Brazil, *see* Brazil
in Ethiopia, *see* Ethiopia
of 1972, map of, 96
of 1972–73, 2
of 1988, 73
of 1988, map of, 78
of 1989, map of, 80
dynamical models, 125–6, 130–1

eastern equine encephalitis, 182
ebola virus, 178
Eguiguren, Victor, 42, 229
El Niño
and climate change, 145
and forecasting, 101–22
and Galápagos Islands, 146–62
and health, 177–88
and hurricanes, 3
and media, 189–201, 210
as natural hazard, 22–6
canonical, 54–6, 84
categories of, 19–22
chronology, 229–36
conditions, diagram of, 70
definition, 2, 19
identification, methods of, 163–73
impacts of 1997–98, 24–5
lessons, 213–16
of 1925, 98
of 1939–41, 98
of 1957–58, 36–37, 167
of 1965, 38
of 1972–73, 39–40, 66, 95–8, 167
of 1976, 99
of 1982–83, 41, 66, 84–100
characteristics of, 85
positive impacts of, 90–4
of 1986–87, 41 99
of 1991–92, 41, 99–100, 111
of 1991–93, 21
of 1991–95, 48, 98
of 1997–98, 41, 66, 71, 100
and media, 191–8
and Galápagos, 146–62
forecasting of, 123–32
phases of, 54–8
surprises, examples of, 206–9
*El Niño Watch*, 197
Eliot, John, 229
encephalitis, 178
eastern equine, 182
Murray Valley, 111, 182
ENSO (El Niño–Southern Oscillation)
and Philippines, 121
and surprise, 202–11
definition of, 2, 18
Environmental News Network, 193
equatorial upwelling, *see* upwelling
Ethiopia Korem Refugee Camp, 180
Ethiopian drought, 106–8
experimental forecasts, 101
models of, 125
Experimental Long-Lead Forecast Bulletin (ELLFB), 125–8

famine in Ethiopia, 106–8
fishery collapse
anchoveta, 36–7, 39, 167–8
sardine (California), 36

fishmeal, 35–6
processing plants, 35, 38
Fleer, Heribert, 134
Flohn, Hermann, 134
floods
of 1988, map of, 79
of 1989, map of, 81
of 1993, 1
Floyd, Hurricane, 51
forecasting
El Niño, 101–22
La Niña, 76–83
1997–98 El Niño, 123–32
forecasts
experimental, 101
hurricane, 47–8
long-lead, 164
of 1991–95, 114–15
of El Niño and La Niña, 7
Fujimori, Alberto, 111–12, 225–8
FUNCEME (Meteorological Organization of Ceará), 109, 118

Galápagos Islands, 146–62
general circulation models (GCMs), 164, 171–2, 187
Gill, Adrian, 234
global
atmospheric GCMs, 172
climate change, 180
warming, 73, 87, 100, 122, 167, 178, 181–2, 187
Godfrey, Stuart, 233
gonadic index, 168
Gray, William, 47
Guano Administration Company, 29, 35–6
guano, 5, 29–32
birds, 5–6, 34, 36, 209
Guayaquil Conference, 222, 232, 236

Hadley Circulation, 138, 144
Halpert, Michael, 72, 76, 93, 235
Hammer, Graeme, 110–11

Hansen, James, 73
hantavirus, 178, 183
Hare, Stephen, 192
Harrison, Edmund, 85
hazard spawners, 27–28
health and El Niño, 177–88
hemorrhagic fever, 178
hepatitis, 184
Hildebrandsson, Hugo, 229
Hoar, Timothy, 235
Holling, C.S., 203
Hoskins, Brian, 234
Houghton, Sir John, 2
Hurricane Andrew, 1, 51
hurricane damage, 23
Hurricane Floyd, 51
hurricane seasonal forecasts, 47–8
hurricanes and El Niño, 3, 93
IMARPE (Peruvian Marine Institute), 38, 46, 231
impacts maps, *see* maps
Indian Ocean, 122
International Decade for Ocean Exploration (IDOE), 232
International Geophysical Year (IGY), 231
International Research Institute (IRI), 119, 135, 235

Jet Propulsion Lab (JPL), 198
jet stream and El Niño, map of, 139

Karoly, David, 234
Kelvin waves, 61–2, 64, 172, 222
Kestin, Tahl, 117, 193, 199–200
Kiladis, George, 64, 68
Kuroshio Current, 62–3
Kyoto Conference of Parties, 193

La Niña
  and media, 189–201
  definition of, 68
  diagram of, 71
  events, chart of, 75
  forecasts, 7, 76–83
  headlines, 68
  indicators, 72–3
  of 1984–85, 76
  of 1998–2000, 68, 71, 73, 76
  Summit, 236
Lake La Niña, 227
Landsea, Chris, 93
Lau, K.-M., 169–70
Lemonick, Michael, 198
limited-area models, 171
Lockyer, Norman, 42
long-lead forecasts, 164

malaria, 166, 178, 182–5, 223
  campaign, stamps of, 179
  in USA, 188
map
  of 1972 droughts, 96
  of 1988 droughts, 78
  of 1988 floods, 79
  of 1989 droughts, 80
  of 1989 floods, 81
  of 1991 droughts, 140
  of 1991 floods, 141
  of 1992 droughts, 142
  of 1992 floods, 143
  of 1997–98 El Niño health impacts, 184
  of 1999 weather anomalies, 82
  of global climate anomalies of 1972, 41
  of Niño regions, 60
maps, Climate Outlook Forum, 136–7
McPhaden, Michael, 123–4
Medécins sans Frontières, 223
media
  and El Niño, 210
  and ENSO events, 189–201
MEI (Multivariate ENSO Index), 67
Mengistu Regime, 180
methods of El Niño identification, 163–73
models
  Cane–Zebiak, 108–9, 205
  coupled, 106
  GCM, 172
  ocean–atmosphere, 164
  dynamical, 125–6, 130
  experimental forecasting, 125
  general circulation, 164, 171–2, 187
  global atmospheric, 172
  limited-area, 171
  statistical, 125–6, 171
mosquitos, 179, 180, 182–5
Mount Pinatubo, 190
Multivariate ENSO Index (MEI), 67
Murphy, Robert Cushman, 36, 230
Murray Valley encephalitis, 111, 182
Myers, Norman, 204

National Meteorological Service
  of Costa Rica, 120
  of Ethiopia, 106–7
  of Kenya, 120
  of South Africa, 118
National Weather Service, 115–16
Nicholls, Neville, 56, 86, 111, 117, 193, 199–200
Niño regions, 60–1

O'Brien, James, 64, 68
outgoing longwave radiation (OLR), 170
overfishing, 92

Pacific Marine Environmental Lab (PMEL), 123
Patz, Jonathan, 188
Peru and 1997–98 El Niño, 225–8
Pezet, Federico Alfonzo, 4, 16
phases of El Niño, 54–8
Philander, George, 73, 92, 120
Philippines and ENSO, 121
Pinatubo, Mount, 190
Piura, Peru, 227
  chart of rainfall, 75
Pizarro, Francisco, 165–6
proxy information, 163–73

Quayle, E.T., 230
Quinn, William, 113, 166, 232
Quinn–Wyrtki forecast, 113

rainfall, La Niña, 77
Rasmusson, Eugene, 85–6, 234
Regional Drought Monitoring Center, 120
remote sensing, 169–70
Rift Valley fever, 178
rodents, 183, 185
Ropelewski, Chester, 72, 76, 235
Rossby waves, 61–3, 172
Rowntree, Peter, 232

Sagan, Carl, 189
sardine fishery collapse (California), 36
satellite images
  of disease risk, 186
  of sea level, 63
Schell, Irving, 231
Schott, Gerhart, 230
sea surface temperature phases, 69
Sears, Alfred, 92, 165–6, 229
Seasonal Climatic Outlook, 116–17, 200–1
septennial rains, 165–6
shigella dysentery, 184
Shukla, J., 43
SOI (Southern Oscillation Index), 45–6, 66, 86, 110–11, 198
  description of, 43
  monthly averages of, 44
South African Weather Bureau, 118
South America Climate Outlook Forum map, 136
Southeast Asia Climate Outlook Forum map, 137
Southern Africa Climate Outlook Forum map, 137
Southern Oscillation, 45, 48, 64, 222
  description of, 43
  diagram, 44
stamps of malaria campaign, 179
statistical models, 125–6, 171
statistics, 164
Steinbeck, John, 36, 231
Stewart, Tom, 203
Stone, Roger, 111
surprise
  and 1991–95 El Niño, 206
  and 1997–98 El Niño, 205–6
  and ENSO events, 202–11
  and media, 210
  examples of El Niño, 206–9
symmetry between ENSO events, 72–3

teleconnections, 3, 133–45, 222
  description of, 47–8
thermocline, 54, 61
  definition of, 53
Thompson, D. F., 204, 210
Todd, Charles, 42, 229
TOGA (Tropical Ocean–Global Atmosphere), 84, 86, 95, 99, 234
TOGA-TAO array, 123, 125, 170, 198, 209
Trenberth, Kevin, 73–4, 233, 235
Tribbia, Joseph, 51
Troup, Sandy, 232
tuberculosis, 178
Turner, Ted, 223
typhoid, 184

UN (United Nations)
  Environment Programme (UNEP), 223
  Foundation for International Partnerships, 223
  General Assembly, 222
upwelling
  coastal, 15, 31–3, 37–8
  equatorial, 54, 58, 60, 71
usable science, 7, 221–4

van Loon, Harry, 64, 68
vector-borne diseases, 182

Walker Circulation, 45, 52–5, 138, 144
  description of, 53
  diagram of, 55
Walker, Sir Gilbert, 42, 45, 47, 118, 164, 230
warm pool, 52–4, 59, 61, 71, 222
weather anomalies of 1999, map of, 82
Webster, Peter, 234
Wooster, Warren, 38
World Bank, 226
World Climate Conference, 181
World Food Conference, 97
World Health Organization, 183, 187
World Meteorological Organization, 222
Wyrtki, Klaus, 59, 103, 113, 233

yellow fever, 178, 185, 187–8

Zebiak, Stephen, 105, 171, 205, 234